Exploring Geography in a
Changing World

2

Simon Ross

DYNAMIC LEARNING

HODDER
EDUCATION
AN HACHETTE UK COMPANY

Hachette UK's policy is to use papers that are natural, renewable and recyclable products and made from wood grown in sustainable forests. The logging and manufacturing processes are expected to conform to the environmental regulations of the country of origin.

Orders: please contact Bookpoint Ltd, 130 Milton Park, Abingdon, Oxon OX14 4SB. Telephone: (44) 01235 827720. Fax: (44) 01235 400454. Lines are open 9.00–5.00, Monday to Saturday, with a 24-hour message answering service. Visit our website at www.hoddereducation.co.uk

© Simon Ross 2009
First published in 2009 by
Hodder Education,
An Hachette UK Company
338 Euston Road
London NW1 3BH

Impression number   5 4 3 2 1
Year   2014 2013 2012 2011 2010 2009

Cover photo: Chad Ehlers/Jupiter Images
Illustrations by Barking Dog Art
Typeset in New Baskerville 11.5/13pt
Layouts by Amanda Easter
Printed in Italy

A catalogue record for this title is available from the British Library

ISBN: 978 0340 946053

Other titles in the series:
Year 7 Pupil's Book: 978 0340 94607 7
Teacher's Resource Book: 978 0340 96973 1
Dynamic Learning Network Edition CD-ROM: 978 0340 94778 4

Year 8 Pupil's Book: 978 0340 94605 3
Teacher's Resource Book: 978 0340 97294 6
Dynamic Learning Network Edition CD-ROM: 978 0340 94779 1

Year 9 Pupil's Book: 978 0340 94606 0
Teacher's Resource Book: 978 0340 97295 3
Dynamic Learning Network Edition CD-ROM: 978 0340 94780 7

CONTENTS

Acknowledgements

I should like to thank the many pupils at Queen's College Taunton who have inadvertently supplied inspiration, feedback and ideas that I have used in the preparation of this book. I would like to thank Ruth Lowe at Hodder Education for all her hard work and her dedication to the pursuit of excellence. I am, as always indebted to my family for their forbearance and support.

In addition I am grateful to the following for supplying information:

Jutta Gehring of BMW, Munich
Mr B. Vher of the Reindeer Rehabilitation Centre, Yokkmokk, Finland
Bryan and Cherry Alexander of Arctic Photos
Sheila Moos of the Danish Sustainability and Wind Alliance
Nick Bullmore of Latitude
Elspetch Goate of the Alpine Centre for Flora and Fauna Education, Zermatt, Austria
Joan Carmichael
RSPCA
Several unnamed but wonderfully helpful people in Lahemaa National Park, Estonia

Dedication

This book is dedicated to Susannah and James.

The Publishers would like to thank the following for permission to reproduce their copyright material:

Photo credits:
p.1 © pintailpictures/Alamy; **p.2** © imagebroker/Alamy; **p.4** *l* Simon Ross, *r* © Colin Underhill/Alamy; **p.7** © Planetobserver/Science Photo Library; **p.9** *tl* © Phil Degginger/Alamy, *bl* © Chris Hellier/Rex Features, *tr* Simon Ross, *br* © Jeff Morgan Cyprus/Alamy; **p.12** © vario images GmbH & Co.KG/Alamy; **p.13** © Kevin Foy/Alamy; **p.14** © Cephas Picture Library/Alamy; **p.21** © Sipa Press/Rex Features; **p.23** *tr* © Bartomeu Amengual/Alamy, *mr* © Cardona Dani/Alamy, *br* © David Robertson/Alamy; **p.24** © Shirley Kilpatrick/Alamy; **p.25** © JAIME REINA/AFP/Getty Images; **p.27** © WILDLIFE/I.Shpilenok/Still Pictures; **p.30** © Mark Bretherton/Alamy; **p.31** © All Canada Photos/Alamy; **p.32** © John Eccles/Alamy; **p.33** © VANDER ZWALM DAN/CORBIS SYGMA; **p.34** *bl* © George Munday/Alamy, *tr* © blickwinkel/Alamy, *br* © Jose B. Ruiz/Nature Picture Library/Rex Features; **p.35** © Valerie Gache/AFP/Getty Images; **p.36** © Stuart Abraham/Alamy; **p.37** *tr* © Gregory Wrona/Alamy, *br* © JUPITERIMAGES/ Agence Images/Alamy; **p.38** © David Chapman/Alamy; **p.39** © THE TRAVEL LIBRARY/Rex Features; **p.41** © Fischer/Alpaca/Andia.fr/Still Pictures; **p.43** *bl* © NASA, Landsat data from USGS EROS Data Center Satellite Systems Branch, earthobservatory.nasa.gov/IOTD/view.php?id=3620, *tr* © Jeff Schmaltz, MODIS Rapid Response Team, NASA/GSFC; **p.44** Simon Ross; **p.45** Simon Ross; **p.46** Simon Ross; **p.47** © StockShot/Alamy; **p.48** © HEKIMIAN JULIEN/CORBIS SYGMA; **p.50** © Spectrum Colour Library/HIP/TopFoto; **p.52** *mr* © Jack Sullivan/Alamy, *br* © blickwinkel/Alamy; **p.53** © Hemis/Alamy; **p.54** *bl* © INTERFOTO Pressebildagentur/Alamy, *tr* Simon Ross; **p.55** © Ken Welsh/Alamy; **p.56** *tr* © Colin Palmer Photography/Alamy, *br* Simon Ross; **p.58** © NASA/GSFC/LaRC/JPL, MISR Team; **p.59** © nagelestock.com/Alamy; **p.60** *tr* © Plinthpics/Alamy, *br* © Jonathan Ball/Alamy; **p.61** © NASA Goddard Space Flight Center; **p.62** © Lina Diksaite; **p.64** © Steve Allen Travel Photography/Alamy; **p.65** *t* © Jason Hawkes/CORBIS, *ml* © ASR Ltd/S Challinor and A Weight/http:// library.coastweb.info/807/1/Microsoft_Word_-_Challinor_and_ Weight_paper.pdf, *mr* © ASR Ltd/S Challinor and A Weight/http:// library.coastweb.info/807/1/Microsoft_Word_-_Challinor_and_Weight _paper.pdf; **p.67** © ROLAND MAGUNIA/AFP/Getty Images; **p.69** *l* © Jon Bower Spain/Alamy, *tr* © PurestockX, *br* © Rex Features; **p.71** © NASA Goddard Space Flight Center (NASA-GSFC); **p.74** © DESIREE MARTIN/AFP/Getty Images; **p.75** © AP Photo/PA Photos/Arturo Rodriguez; **p.76** Simon Ross; **p.77** © thislife pictures/Alamy; **p.78** © Cultura/Alamy; **p.79** © Michael DeFreitas Europe/Alamy; **p.81** © Andrei Nekrassov/Alamy; **p.82** *both* Simon Ross **p.85** © Sari Gustafsson/Rex Features; **p.86** *tl* Simon Ross, *tr* © Ian M Butterfield/Alamy; **p.88** © SIR Salzburger Institut für Raumordnung & Wohnen, Inge Strassl; **p.89** *both* © DEMEKAV (Volos Municipal Enterprise for Urban Studies, Construction and Development); **p.90** © Photodisc/Getty Images; **p.91** © Brendon Bishop/Alamy; **p.93** © Graham Watts/Alamy; **p.95** *tr* © Poula Hansen/istockphoto.com, *tm* Simon Ross, *tl* © David R. Frazier Photolibrary, Inc./Alamy; **p.96** *t* © travelib prime/Alamy, *b* © Gail Johnson - Fotolia.com; **p.99** © Bryan & Cherry Alexander Photography; **p.100** © Bryan & Cherry Alexander Photography; **p.101** © imagebroker/Alamy; **p.103** *t* © vario images GmbH & Co.KG/Alamy, *b* © mmadpic/Alamy; **p.105** © AFP PHOTO/GETTY IMAGES/HO; **p.106** © Jochen Tack/Alamy; **p.107** © Neil lee Sharp/Alamy; **p.108** *t* © Marcel Mochet AFP/Getty Images, *br* © Soren Andersson/AFP/Getty Images; **p.110** © Skyscan Photolibrary/Alamy; **p.112** © James L. Amos/CORBIS; **p.113** ©

Bartek Wrzesniowski/Alamy **p.114** © 2000 BMW AG, Munchen, Deutschland. All rights reserved; **p.115** © 2000 BMW AG, Munchen, Deutschland. All rights reserved; **p.118** © Tony West/Alamy; **p.119** © PCL/Alamy; **p.120** *tr* © David Pollack/K.J. Historical/Corbis, *bl* © Photodisc/Getty Images, *bm* © Photodisc/Getty Images, *br* © Moritz von Hacht/iStockphoto.com; **p.122** *bl* © Peter Titmuss/Alamy, *br* © Peter Salton/Alamy; **p.123** *l* © Gregory Wrona/Alamy, *r* © Peter Turnley/CORBIS; **p.124** *l* © Greg Balfour Evans/Alamy, *r* © Nordicphotos/Alamy; **p.125** © Rob Cole Photography; **p.127** © Robert Harding/Rex Features; **p.128** © Ta' Mena Agri Ltd.; **p.129** © Chris Howes/Wild Places Photography/Alamy; **p.130** © LOOK Die Bildagentur der Fotografen GmbH/Alamy; **p.131** © Frank Schroeter/Visum/Still Pictures; **p.132** *t* © Eric James/Alamy, *m* © Alexander Nemenov/AFP/Getty Images, *b* © PASCAL GUYOT/AFP/Getty Images; **p.135** *t* © Juan Carlos Calvin/age fotostock/ Photolibrary.com, *b* © PhotoStock-Israel/Alamy; **p.136** © Biosphoto/Astruc Lionel/Still Pictures; **p.137** *bl* © Plan Bleu, *br* Reproduced with kind permission of European Ecolabel; **p.138** *l* © tbkmedia.de/Alamy, *tr* © Rod Edwards/Alamy, *br* © Nigel Hicks/Alamy; **p.139** © imagebroker/Alamy; **p.140** *both* Simon Ross; **p.142** © Mike Greenslade/Alamy.

Text and image acknowledgements:
Maps on the cover flaps Octopus Publishing Group: *Philip's Modern School Atlas*, © 2007 Philip's; **p.16** Met Office website © Crown Copyright; **p.18, 19** Climate weather data for London and Berlin, from *World Weather Book*, 5th Revised Edition (Hutchinson, 2000); **p.24** Graph showing growth in tourists visiting Mallorca, from 'Tourism Development in Mallorca: Is Water Supply a Constraint? by Stephen Essex, Martin Kent and Rewi Newnham from *Journal of Sustainable Tourism*, 1/1/04 © Taylor & Francis, reprinted by permission of the publisher (Taylor & Francis Group, http://www.informaworld.com); **p.32** Map extract of Pyhä-Luosto National Park © Karttakeskus / Affecto Finland Ltd, Finland, 2009; **p.40** Octopus Publishing Group: *Philip's Modern School Atlas*, © 2007 Philip's; **p.42** Map extract of Entrèves © IGN PARIS 2003; **p.50** Map extract of Sicily © Rough Guides; **p.63** Forbidden activities information © Direction Kursiu Nerija National Park; **p.72** Information about Sweden, top © European Communities, 1995-2009; Information about Sweden, bottom "Sweden" as found at http:// www.infoplease.com/ipa/A0108008.html as it appeared on Information Please® Database, © 2007 Pearson Education, Inc. All rights reserved; **p.73** Population data © Eurostat; **p.77** Roseberry Mews extract, © Woodford Group PLC, 2007; **p.83** Map extract of Tallinn © Jāna sēta Map Publishers 2007; **p.92** © NPR News; **p.98** Reproduced by permission of Ordnance Survey on behalf of HMSO. © Crown copyright 2009. All rights reserved. Ordnance Survey Licence number 100036470; **p.104** Freedom Food screenshot © RSPCA 2009; **p.109** Data on electricity production © Eurostat; **p.111** Reproduced by permission of Ordnance Survey on behalf of HMSO. © Crown copyright 2009. All rights reserved. Ordnance Survey Licence number 100036470; **p.115** Copyright 2000 BMW AG, Munchen, Deutschland. All rights reserved; **p.116** Data © Europe's Energy Portal; **p.126** Michelin et Cie, 2009, Authorisation No. GB0904003. Extract from ZOOM map 106 – Environs of Paris; **p.130** Sustainable tourism checklist © Crown Copyright; **p.141** Map extract of Lahemaa National Park © Regio maps.

Every effort has been made to trace all copyright holders, but if any have been inadvertently overlooked, the Publishers will be pleased to make the necessary arrangements at the first opportunity.

Introduction

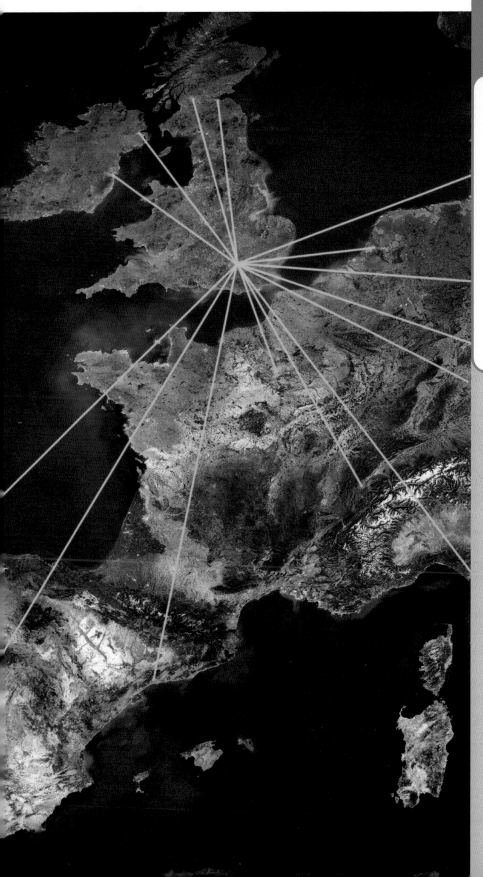

In this chapter you will study:

- making connections with Europe
- where Europe is in the World
- what makes Europe special
- European landscapes
- European Union
- should Turkey join the EU?

Making connections with Europe

In Book 1 we focused on the UK. In this book we will be focusing on Europe, particularly the European Union.

Did you know that Britain, Ireland and France have not always been separated by the sea? Thousands of years ago we were all once joined together by land, and we were once physically part of Europe. As sea levels rose at the end of the last ice advance, we then became separate and our islands formed.

Nowadays, we are getting much more involved with the rest of Europe again. School trips frequently take pupils into Europe, such as skiing in the Alps (Figure 1). Many school children take part in French, German or Spanish exchanges.

▲ Figure 1 Children skiing in the Alps

Activities

1 Study Figure 1.
   a) Why do you think schools run trips to the Alps?
   b) What school trips take place from your school to Europe?
   c) Which European countries have you visited?
   d) For one of the countries you have visited, describe what you did and what you liked about the country.

2 Study Figure 2. It shows the European countries visited most recently by members of a class of Year 8 pupils.
   a) Carry out a survey to find which European countries the members of your class have visited most recently. (Alternatively, consider all European countries visited by each member of the class.)
   b) Present your results as a flow map similar to Figure 2 or a bar chart.
   c) Which country is the most visited by your class?
   d) Are the countries closer to the UK visited more often than those further away?
   e) Try to suggest reasons why some countries are visited more often than others.

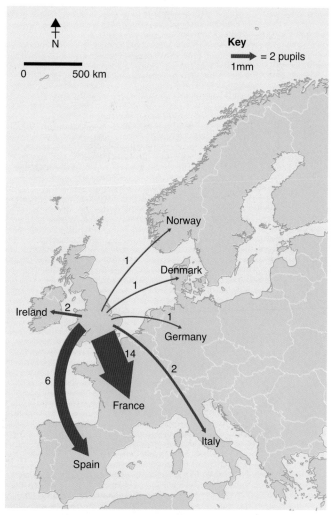

▲ Figure 2 Flow diagram showing visits of a Year 8 class to European countries

It has never been easier to connect with the rest of Europe by road, rail and air.

For many years people have travelled from the UK in boats to connect with Europe. Today, ferries still carry many people and cars across the English Channel, the North Sea and the Irish Sea to Europe.

In 1994, the Channel Tunnel was opened as a link with France, and so to the rest of Europe. It is a tunnel under the English Channel, from Folkestone in the UK to Calais in France. Today, up to three times an hour, fast trains regularly carry people and vehicles through the 50-kilometre tunnel in just 35 minutes (Figure 3). It takes less time than the ferries and is not interrupted by bad weather.

Activity

3 Study Figure 3.
   a) Look at the map. Why do you think the Channel Tunnel was constructed at this point across the English Channel?
   b) The Channel Tunnel is not just a single tunnel. How many tunnels are there?
   c) What do you think is the purpose of the service tunnel?
   d) Piston relief ducts have been constructed at every 250 m along the tunnel. What do you think their function is?
   e) Why do you think some people do not wish to travel through the Channel Tunnel?
   f) If you and your family were going to travel to France, would you prefer to cross the English Channel by ferry or by train through the Channel Tunnel? Give reasons for your answer.

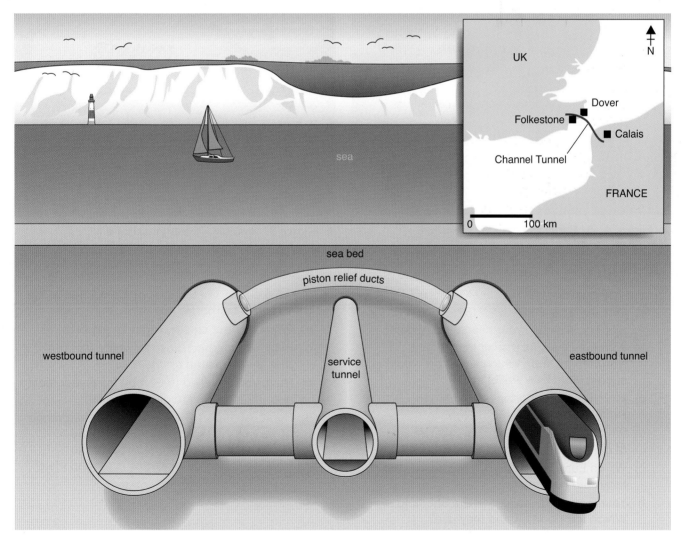

▲ **Figure 3 The Channel Tunnel**

The high-speed inter-continental Eurostar rail service also makes use of the Channel Tunnel. In 2007, the new Eurostar terminal at London St Pancras was opened (Figure 4). Journeys to Paris from London now take just over 2 hours.

Cheap airline flights using carriers such as easyJet, Ryanair and Flybe enable people to travel to many parts of Europe from regional airports throughout the UK (Figure 5). Whilst these cheap flights are popular, there is increasing concern about pollution from the aeroplanes and their contribution to carbon emissions.

The growth of electronic communications, particularly the internet, has led to a revolution in connecting people in the UK with individuals and organisations in other countries. It is very easy to talk with people in Europe, buy products from European shops and make bookings at hotels and restaurants. Satellite television enables us to watch programmes broadcast from Europe.

▲ Figure 4 St Pancras Eurostar terminal

▲ Figure 5 Ryanair – a budget airline

Activities

4 Study Figure 6.
   a) You and your family wish to travel to France by ferry. Which is your nearest ferry port?
   b) Your family wants to visit Ireland (not Northern Ireland). Which ferry port would you suggest that they travel from?
   c) Which UK ferry port is connected to Rotterdam?
   d) Which UK ferry port has the greatest number of connections with mainland Europe?
   e) Which European countries are connected by ferry with Newcastle?
   f) Your family wishes to visit the Shetlands. Which ferry port should they use?

5 Figure 6 shows the location of a selection of the UK's regional airports.
   a) Where is your closest regional airport?
   b) Is there an airport close to where you live that is not located on the map?
   c) Have you travelled to Europe from your local airport? If so, where did you travel?
   d) Regional airports have expanded their range of services in recent years, taking the pressure off international airports like Heathrow. Do you think this is a good idea? Explain your answer.

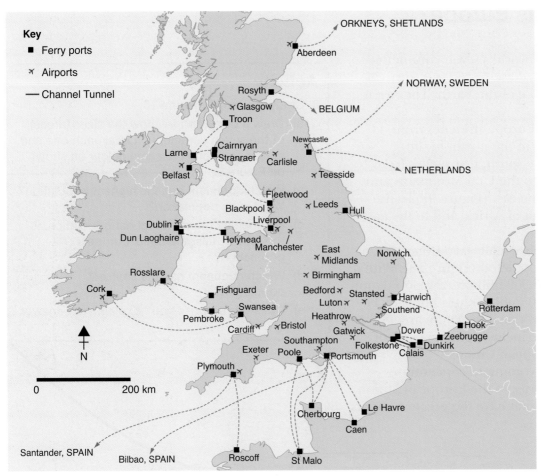

▲ **Figure 6   Map showing UK connections with Europe by ferry**

My name is Shelia. I want to travel by Eurostar from London St Pancras to Paris. I want to leave London on Friday next week in the afternoon and return on Saturday in the afternoon. Can you find me the cheapest tickets and give me the train times? Thanks.

www.eurostar.com

Hello. How much is the cheapest fare for a day return ticket (1 car without a trailer and 4 adult passengers) from Dover to Calais by ferry on the first day of next month? Can you give me the sailing times and the cost? Cheers. BB

www.seafrance.com/seafrance/opencms/uk/en/passenger

Hi, I'm Mandy. Can you fly me from Bristol to Prague on easyJet anytime next Wednesday or Thursday? I just need a single adult ticket as I am meeting my boyfriend. Can you give me some times and prices? Thanks.

www.easyjet.com/en/book/index.asp

▲ **Figure 7   European travel queries**

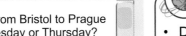

ICT ACTIVITY

- Do some research on travelling to Europe using the ferry, Eurotunnel, easyJet and Eurostar.
- Look at Figure 7. Imagine that you are an online travel adviser. Three people have requested help with arranging travel to Europe. Select one or more of the queries and try to help by using the internet. The relevant website is next to each of the queries. Present each answer in the form of an e-mail.

Where is Europe?

Europe is one of the world's seven continents (Figure 8). Together these continents form 29 per cent of the surface of the Earth. The rest is water!

The continent of Europe stretches from Iceland and Finland in the north, to the Mediterranean in the south. It stretches from Ireland in the west, to the Ural Mountains and the Caspian Sea in the east (Figure 9). To the south is the continent of Africa, and to the east the continent of Asia.

Europe is one of the smallest continents yet it is one of the most populated (Figure 10). With a population of over 7,000 million (the population of the UK is about 60 million), it is home to 11 per cent of the world's people. There are 48 countries in Europe (including Russia), 27 of which form the **European Union**.

In this book we are going to concentrate mainly on the countries of the European Union, as they are the ones that have the greatest impact on our lives in the UK.

Activity

6 Study the data in Figure 10.

   a) Draw a bar graph to show the sizes (areas) of the seven continents. Label each one and colour each one the same colour as shown on Figure 8.

   b) Which continent has the highest population? What is its population?

   c) Europe has the third-highest population. Is this true or false? What is Europe's population?

   d) Which continent has the smallest population?

   e) Does this surprise you? Explain your answer.

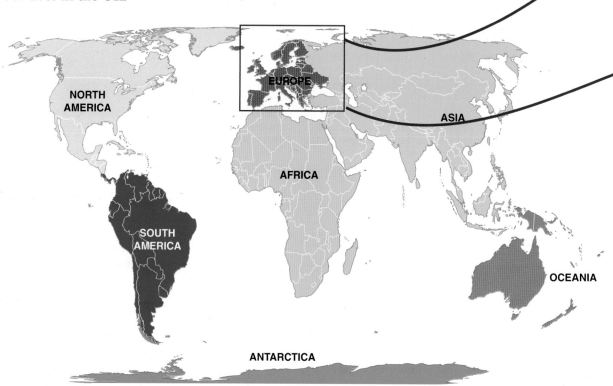

▲ Figure 8 World map showing continents

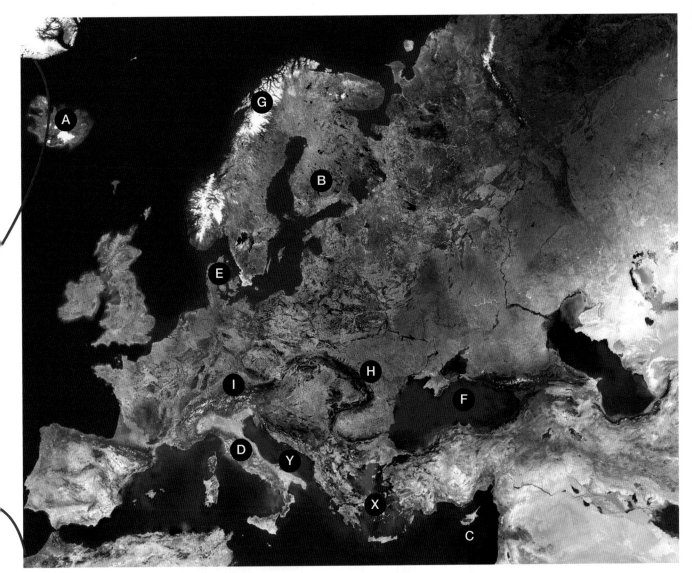

▲ Figure 9   Satellite view of Europe

Continent	Area (million sq km)	Approx. population in millions (2002)	Percent of total population
Asia	45.0	3,800	60
Africa	30.3	890	14
North America	24.7	515	8
South America	17.8	371	6
Antarctica	12.1	0.001	0.00002
Europe	9.9	710	11
Oceania	8.9	33.6	0.06

▲ Figure 10   World continents (area and population)

Activity

7 Study Figure 9 and the atlas maps on the inside front and back covers.
   a) Name the European countries A–E.
   b) What is the name of the sea at F?
   c) Can you explain the white areas on the photo at G?
   d) What are the names of the mountain ranges at H and I?
   e) Is the Aegean Sea at X or Y?
   f) Notice that most of northern Europe is green, whereas Southern Europe is light brown. How do you account for this?

What makes Europe special?

Consider what it is to be European. How does Europe differ from other continents? Look at Figure 11. It shows what a group of pupils in Year 8 think Europe is like. Their views or **perceptions** are the result of their travels or what they have seen or read about. Notice how many views are based on famous places or buildings.

Activities

8 Conduct a survey in your class to find the views of your classmates about what Europe means to them. What do they think being 'European' means? What do they associate with 'Europe'? Present your information as a wall display using words, sketches and photos.

9 Choose another continent (for example, Africa). How do you think life in your chosen continent differs from life in Europe? Think about the weather, the lifestyle, the language and other differences.

▲ Figure 11

European landscapes

Europe has an incredible range of physical landscapes (Figure 12). In Iceland there are huge icecaps, majestic glaciers and vast areas of wilderness (Figure 12a). Scandinavia also has glaciers as well as dramatic fast-flowing rivers, beautiful forests and lakes.

There are several impressive mountain ranges in Europe including the Alps (Figure 12b), the Pyrenees and the Carpathians. Much of northern Europe is low-lying, favouring the development of transport communications and settlements. Much of this land is used for farming. Some parts of Europe are low marshes, such as the Camargue in France (Figure 12c).

Parts of Southern Europe have a very dry climate and a dusty, semi-arid landscape with few plants (Figure 12d). This landscape is very different from regions further north.

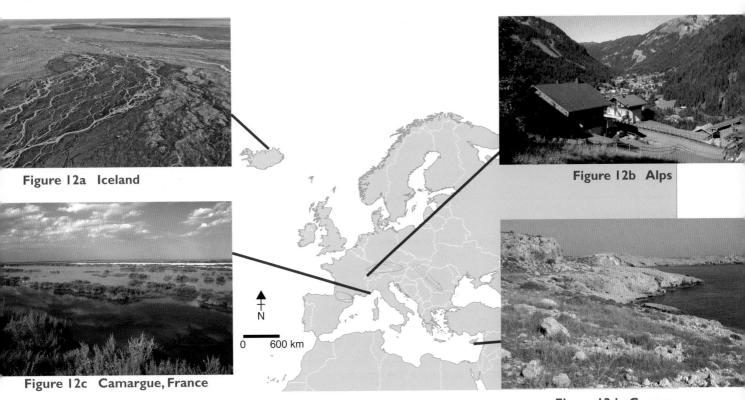

Figure 12a Iceland

Figure 12b Alps

Figure 12c Camargue, France

Figure 12d Cyprus

▲ Figure 12 Landscapes in Europe

Activities

10 Study Figure 12.
   a) Make a list of the different landscape features that you can see in the photographs.
   b) Which photograph do you like best and why?
   c) Would you be interested in visiting the place where your chosen photograph was taken? Why?
   d) Do you think any of the photographs show scary or dangerous places? Explain your answer.

11 Turn to the atlas maps in the inside front and back covers. Imagine that you could travel to any place in Europe (that you haven't already visited) to take a photograph of the physical landscape.
   a) Where would you choose to go and why?
   b) What do you think the landscape would look like there?
   c) Use the internet to try to find a photograph of this area. Is the photograph what you expected your chosen place to look like?
   d) What do you like about the photograph?

The European Union

The European Union (known as the EU) is a group of countries whose governments work together. It is rather like a club. To join the 'club', countries agree to follow certain rules and in return they receive benefits and support. All countries pay into a central fund. Mostly this comes from the taxes that people pay. The European Union uses most of this money to help people in its member countries and to improve the quality of peoples' lives. Figure 13 describes some of the functions of the European Union.

The European Union started in 1957 with just six countries (Figure 14). In 2007, the club increased to a membership of 27 countries. Other countries are likely to join in the future.

Four countries in Europe belong to another organisation called the **European Free Trade Association** (EFTA). This was set up in 1960 at about the same time as the European Union started. Its member countries (Iceland, Liechtenstein, Norway and Switzerland) have arranged **free trading** agreements with countries elsewhere in the world.

- Brings people in Europe closer together.
- Makes it easier for Europeans to buy and sell things to each other by encouraging free trade between countries.
- Removes controls that stop Europeans moving around freely within the EU.
- Encourages exchange visits that help ordinary Europeans to understand what they all have in common.
- Supports farming and food production and tries to keep food prices low for shoppers.
- Encourages justice and peace throughout the world.
- Produces guidelines on a range of products throughout the EU.
- Supports people who live in the poorer parts of the EU.

▲ **Figure 13 What does the EU do?**

Activities

12 Study Figure 14. Draw a timeline to show when the countries joined the European Union.

  a) To do this you need to draw a line to scale to represent the 50 years between 1957 and 2007. A scale of 4 cm = 10 years would work well, giving you a total timeline of 20 cm. Draw your timeline in pencil.

  b) Now use the information in Figure 14 to locate on the timeline the dates when the countries joined the European Union. Use arrows and write the country names alongside.

  c) Locate the year of your birth on the timeline.

  d) Now complete your diagram by writing a title and a scale.

13 You will need a blank European outline map, which your teacher can give you.

  a) Use a colour of your choice to shade the 27 countries that belong to the European Union (Figure 14).

  b) Write the name of each country onto your map.

  c) Use a different colour to shade Turkey, which hopes to join the European Union in the future. Write the name of the country on your map.

  d) Use a third colour to shade the four countries that belong to the EFTA (see the text above).

  e) Complete your map by adding a title 'Europe's Economic Clubs'. Explain the three colours in a key.

1957	Treaty of Europe establishes the European Economic Community (now called the European Union) with six countries – France, West Germany, Netherlands, Belgium, Luxembourg and Italy.
1973	UK, Ireland and Denmark join the EC.
1981	Greece becomes the 10th member.
1986	Spain and Portugal join in.
1995	Austria, Finland and Sweden join in. Norway refuses to join after holding a referendum (national vote) on the issue.
2002	Euro notes and coins are introduced. Of the 15 countries, Sweden, Denmark and the UK decide to keep their existing national currencies.
2003	Ten new countries join the EU – Cyprus, the Czech Republic, Estonia, Hungary, Poland, Slovenia, Slovakia, Latvia, Lithuania and Malta.
2007	Bulgaria and Romania join the EU, bringing the total to 27 countries.

▲ **Figure 14   Growth of the European Union**

Issue: Should Turkey join the EU?

Turkey may well be the next country to join the European Union. However, not all European countries agree that Turkey should be allowed to join. They have concerns about Turkey's refusal to recognise Cyprus, a country divided between Greek and Turkish sectors. There are also concerns about human rights issues in Turkey. Some people are worried about the impact of allowing a relatively poor country with a growing population to join the European Union.

Activity

14 Find the location of Turkey on the map in the inside back cover, then study Figure 15. It contains comments submitted by people to a website.

a) Read through the comments. What are the main arguments against Turkey joining the European Union?

b) Why do some people believe that Turkey should be allowed to join the European Union?

Comments

How is Turkey considered part of Europe? What country will join the EU next, Israel? The EU is too big already, Turkey shouldn't be allowed to become a member. **Anthony Cosgrove, London, England**

Turkey has worked really hard to meet EU standards and the EU should let it join. If we didn't want Turkey to join, we should have said so from the start. **Lucas Hernu, Orlean, France**

It would be a scandal if the European Union didn't let Turkey join! Surely if any country wants to join and they meet the EU standards they should be allowed to be a member. **Jim Dancy, Edinburgh, Scotland**

Absolutely not! Why should Turkey be allowed to join the EU? Turkey is just not part of Europe. **F. Jergen, Netherlands**

If Turkey *is* allowed to become a full member, millions of Turks will move to live in other countries within the EU. The EU is not ready for that to happen. The cultural difference is too wide. **Abdi, Istanbul, Turkey**

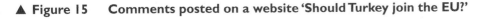

▲ Figure 15   Comments posted on a website 'Should Turkey join the EU?'

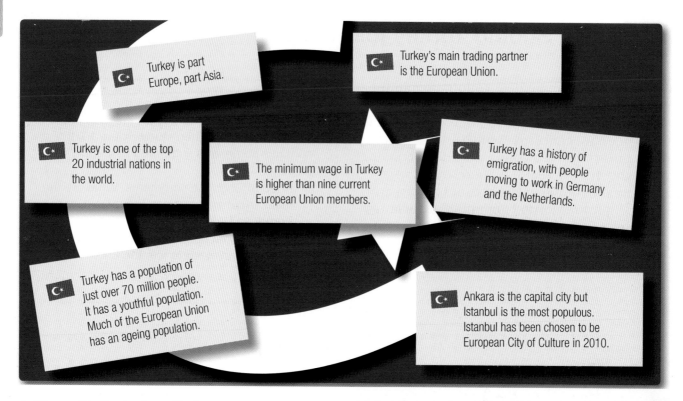

▲ **Figure 16   Facts about Turkey**

The facts about Turkey shown in Figure 16:

Turkey is part Europe, part Asia.

Turkey's main trading partner is the European Union.

Turkey is one of the top 20 industrial nations in the world.

The minimum wage in Turkey is higher than nine current European Union members.

Turkey has a history of emigration, with people moving to work in Germany and the Netherlands.

Turkey has a population of just over 70 million people. It has a youthful population. Much of the European Union has an ageing population.

Ankara is the capital city but Istanbul is the most populous. Istanbul has been chosen to be European City of Culture in 2010.

Activity

15 Study Figure 16.

   a) Find out about the geography of Turkey, using the internet to help you. Find a map of Turkey and highlight the major cities. Discover which countries border Turkey.

   b) Try to discover some information about the population, the economy and the natural environment of Turkey. Find out why Turkey is such a popular holiday destination with people from the UK.

   c) Create a report that presents the information you have found on Turkey (using a mixture of writing and pictures.

   d) Do you think Turkey should be part of the European Union? Why?

Some useful websites

www.allaboutturkey.com/cografya.htm

http://geography.about.com/library/maps/blturkey.htm

www.infoplease.com/ipa/A0108054.html

▲ **Figure 17   Istanbul**

Weather and Climate

In this chapter you will study:

- the weather in Europe
- patterns of temperature
- the climate of Europe
- extreme weather: the European heatwave (2003)
- the water supply in Mallorca, Spain.

A   What is the weather?

The **weather** is the day-to-day condition of the atmosphere. When we describe the weather we talk about how warm it is, whether the sun is shining or if it is going to rain.

The weather affects all our lives a great deal (as you discovered in Book 1). If the weather is warm and sunny it encourages us to spend time outside. We buy and wear different clothes depending on the weather (such as T-shirts and shorts, or jumpers and coats). Have you noticed how shops stock clothes for different seasons? In warm weather we spend money in shops buying drinks, ice cream and salad vegetables (Figure 1). Did you know that supermarket managers are very watchful of the weather, as it affects what they choose to stock their shelves with?

▲ **Figure I   French supermarket**

Activities

1 Study Figure 1.

  a) What is the evidence that this supermarket is selling food for use in warm summer weather?

  b) What other items would you expect to find in this supermarket specifically aimed for warm and sunny weather?

  c) What suggestions would you make to the supermarket manager for stocking the shelves for cold winter weather?

  d) If you went into a high-street clothes shop, what items would you expect to find displayed for autumn weather?

2 Suggest why the following groups of people and organisations keep a careful eye on the weather:

  a) ferry companies

  b) local council road gritters

  c) airport authorities

  d) festival organisers (e.g. Glastonbury)

  e) farmers.

ICT ACTIVITY

Mr Jones is the fresh food buyer for a large supermarket in your local area. He needs to place an order with his supplier for fresh fruit and vegetables for next week. Help him to decide what to order.

- Find out next week's weather forecast by looking at sites such as the Met Office (www.metoffice.gov.uk/weather/uk/uk_forecast_weather.html) or Met Check (www.metcheck.com). Write a few sentences describing the forecast. Is the weather expected to be different from that of this week?

- In the light of next week's forecast, suggest any items of fresh fruit or vegetables that you think should be bought by Mr Jones.

- Mr Jones' colleague Mrs Stoat is the senior buyer for the supermarket. Her job is to oversee the ordering in all departments. Can you offer her any suggestions for next week? For example, if it is due to be wet, you might suggest that she has plenty of umbrellas for sale!

B Describing patterns of temperature in Europe

Look at Figure 2. It shows the pattern of temperatures across Western Europe on Sunday 23 December 2007 at 10 am. Look at the key to see that the blue colours represent areas with temperatures below 0°C. The warmest areas on the map are shaded with orange and red colours.

When looking at **patterns** on a map, we are looking at the overall trends. Where are the high values and low values? Do the values change gradually or rapidly in different parts of the map? Is there a trend of change in one particular direction?

Look for the following patterns of temperature on Figure 2.

● Temperatures get colder towards the north and east of Europe.

● The warmest temperatures are in the south of Europe and in North Africa.

● In Portugal and Spain, the warmest temperatures are along the coastline.

● In northern Europe it is warmer at the coast than inland.

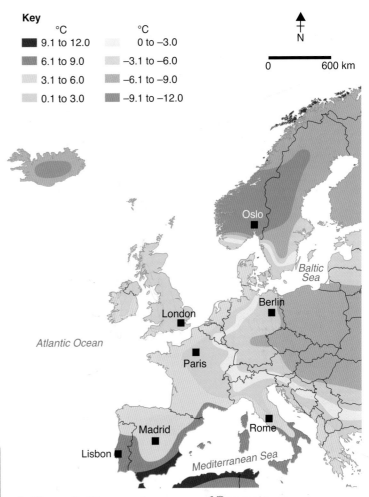

Key

°C	°C
9.1 to 12.0	0 to −3.0
6.1 to 9.0	−3.1 to −6.0
3.1 to 6.0	−6.1 to −9.0
0.1 to 3.0	−9.1 to −12.0

0   600 km

▲ Figure 2 Temperature map of Europe

Activity

3 Study Figure 2.

   a) Make a copy of the table in Figure 3.

   b) Locate each of the cities on Figure 2 and place a tick in the correct column to show the temperature value.

   c) Which city was coldest on 23 December 2007?

   d) Which city was the warmest?

   e) Why do you think many people from the UK choose to travel to southern Europe and North Africa over Christmas?

   f) Why do some people from the UK choose to travel to eastern and northeastern Europe over Christmas?

   g) Some people from the UK choose to retire to the Mediterranean coast of France or Spain. How does the information in Figure 2 help to explain this?

   h) If you could travel to somewhere in Western Europe for the Christmas period where would you go and why?

Temperatures (°C)							
	−9.1 to −12	−6.1 to −9	−3.1 to −6	−0.1 to −3	0 to 3	3.1 to 6	6.1 to 9	9.1 to 12
Oslo								
Berlin								
London								
Paris								
Madrid								
Lisbon								
Rome								

▲ Figure 3 European temperatures for selected cities (23 December 2007)

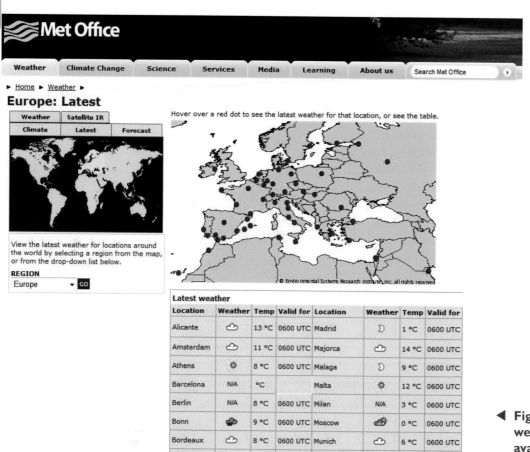

◀ **Figure 4 European weather information available from the Met Office**

ICT ACTIVITY

Up-to-date information about the weather in Europe can be accessed at the Met Office's website at

www.metoffice.gov.uk/weather/europe/europelatest. html (Figure 4).

- Using a blank outline map of Europe, mark the location of a selection of European cities with a pencil. Write the name of each city alongside. Aim to have 10 to 15 cities in total.

- For each of your chosen cities, look at the weather for today on the internet, and then show the weather information using the symbols given on the website. Write the temperature value alongside each symbol. Use colours to make the map more attractive.

- Write a short weather report (to be read out to the rest of the class) in the style of the weather forecasts that you see on the TV or hear on the radio.

C Explaining patterns of temperature in Europe

Figure 2 on page 15 shows that it is colder in northeast Europe than in southern Europe. This is typically the situation in the winter. One of the main reasons for this pattern is the power of the sun at this time of the year.

Temperatures on the Earth's surface are affected by the different positions of the sun in the sky. Look at Figure 5, which shows a torch beam shining on a curved surface. This is rather like the sun shining on the curved surface of the Earth. Notice that the beam of light is very concentrated at **A** where the beam shines directly down on the surface. At **B** where the beam is at an angle, the light is less strong and is more spread out on the surface. This is similar to the way that heat from the sun warms the Earth.

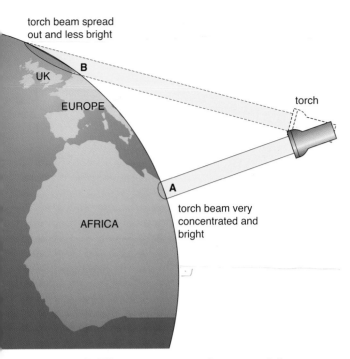

torch beam spread out and less bright

B

UK

EUROPE

torch

A

torch beam very concentrated and bright

AFRICA

▲ **Figure 5 The sun compared to a torch beam**

Southern Europe has a warmer climate than northern Europe because it is closer to the **Equator**. This means it receives more concentrated heat from the sun than areas closer to the Poles (you will learn more about this in Book 3). In December (see Figure 2) the sun is still quite strong in southern Europe (even though it is winter). In the north the sun is very weak, and in the far north it doesn't shine at all for several months. Imagine living in 24 hours of darkness!

Another pattern shown on Figure 2 is that temperatures tend to be higher at the coast than inland. This is because the sea loses heat less quickly than the land in the winter and so it keeps the coastal areas warmer than further inland. In addition, an ocean current called the **North Atlantic Drift** warms the west coast of northern Europe (Figure 6). This major ocean current spreads heat northwards from the Caribbean towards the UK and Norway, bringing with it mild weather in the winter.

Activities

4 Study Figure 5. Remember that the torch represents the sun, and the torch beam represents the sun's rays.

  a) Make a careful copy of Figure 5 using a yellow colour to show the torch beam.

  b) At position **A**, when the torch is directly overhead, the light is more concentrated. At position **B**, when the torch is at a lower angle to the surface, the light is not as bright. Try to explain why.

  c) Write a few sentences describing how the ideas shown in Figure 5 can help explain why southern Europe tends to have higher temperatures than northern Europe.

5 Study Figure 6.

  a) What is the name of the ocean current that brings warm conditions to northwestern Europe?

  b) Where does the ocean current come from?

  c) If the ocean current did not exist, how do you think the temperature map (Figure 2 on page 15) would be different?

  d) Some scientists have suggested that one effect of global warming might be to reduce the effectiveness of the North Atlantic Drift. If this were to happen, what impact would this have on people's lives in northwestern Europe?

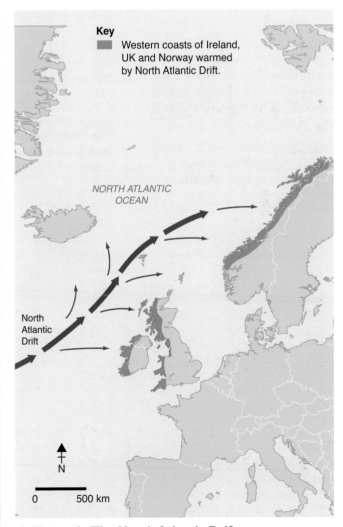

Key

  Western coasts of Ireland, UK and Norway warmed by North Atlantic Drift.

NORTH ATLANTIC OCEAN

North Atlantic Drift

N

0   500 km

▲ **Figure 6 The North Atlantic Drift**

D Climate of Europe

The **climate** of a place is its average weather, usually worked out over a period of 30 years. It differs from the weather, which describes the state of the atmosphere over a matter of a few days.

Look at Figure 7. It shows a selection of climate information about London. Compare the average temperatures with the highest and lowest temperatures. Remember that an average figure will always be somewhere in between the extremes so it will never seem terribly dramatic. In not reflecting the extreme values, averages can be a bit misleading although they are widely used in comparing sets of data.

Climate data is often represented in the form of a graph called a **climate graph** (see Book 1). Figure 8 shows the climate graph for London. Notice that bars show rainfall whereas temperatures are shown using line graphs.

	Sunshine (average hours per day)	Temperatures (°C)				Precipitation (mm)	
		Average daily		Highest recorded	Lowest recorded	Average monthly	Wet days (more than 0.25 mm)
		minimum	maximum				
Jan	1	2	6	14	-10	54	15
Feb	2	2	7	16	-9	40	13
Mar	4	3	10	21	-8	37	11
April	5	6	13	26	-2	37	12
May	6	8	17	30	-1	46	12
June	7	12	20	33	5	45	11
July	6	14	22	34	7	57	12
Aug	6	13	21	33	6	59	11
Sept	5	11	19	30	3	49	13
Oct	3	8	14	26	-4	57	13
Nov	2	5	10	19	-5	64	15
Dec	1	4	7	15	-7	48	15

▲ **Figure 7 Table of climate information for London**

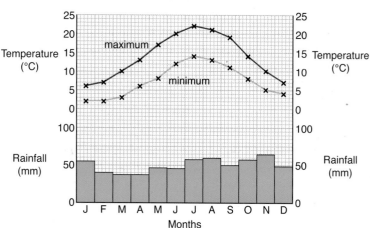

▲ **Figure 8 Climate graph for London**

Activity

6 Study Figure 7.

a) What is the average daily maximum temperature for London in this current month?

b) How many hours of sunshine per day would you expect this month?

c) How many days would you expect it to be wet this month?

d) Which is the hottest month in London, July or August?

e) Which month is the coldest, January or February?

f) A favourite aunt from abroad wants to visit London in October. What clothes would you advise her to bring and why?

g) A lot of people like to visit London in May. How do you think the climate in May helps to explain this?

	Sunshine (average hours per day)	Temperatures (°c)				Precipitation (mm)	
		Average daily		Highest recorded	Lowest recorded	Average monthly	Wet days (more than 0.25 mm)
		minimum	maximum				
Jan	2	-3	2	13	-21	46	17
Feb	2	-3	3	17	-22	40	15
Mar	5	0	8	22	-14	33	12
April	6	4	13	30	-6	42	13
May	8	8	19	32	-3	49	12
June	8	12	22	35	3	65	13
July	8	14	24	37	5	73	14
Aug	7	13	23	37	6	69	14
Sept	6	10	20	34	1	48	12
Oct	4	6	13	25	-4	49	14
Nov	2	2	7	17	-9	46	16
Dec	1	-1	3	15	-18	43	15

ICT ACTIVITY

Find out about the climate of a European country of your choice using the Met Office's website www.metoffice.gov.uk/weather/europe/europepast.html. How do you think the climate affects people's lives in your chosen country? Search for some photographs to illustrate your account.

◀ **Figure 9 Table of climate information for Berlin**

Activities

7 Study Figure 9, which contains data for the climate of Berlin.

a) Draw a climate graph for Berlin. Use Figure 8 to help you draw the axes and get started.

b) Work in pairs to suggest appropriate positions on your graph to add the following letters/labels to create a 'living graph'.

A Franz dressed warmly today for open-air ice-skating in the park. He decided not to wear his baseball cap due to the lack of sunshine.

B Natasha decided not to go outside today, as it was far too cold.

C It was baking hot today as Johanna slapped on sun cream and put on a hat to go to the local shop to buy an ice-cool drink. Temperatures soared to 35°C.

D Yet another wet miserable day as Emil, Ingrid and Stefan prepared themselves for cross-country running training. They wore old trainers on account of the muddy paths they expected to be running on.

c) Write and position on the graph two more 'living graph' statements.

8 Look at Figure 10 on page 20. It shows the different climates of Europe.

a) Make a copy of the map in Figure 10 on a blank outline map of Europe. Don't forget to complete the key.

b) Locate Spain on Figure 10. Describe the typical climatic conditions experienced by people living in Spain.

c) People living in the Baltic countries of Lithuania, Latvia and Estonia experience a continental climate. How would their summers compare with summers in Spain?

d) Suggest some advantages and disadvantages of living in the Alps and experiencing a mountain climate.

e) People sometimes say that the climate of the UK is 'boring'. Do you agree with this sentiment? Use information in Figure 10 to support your argument.

Key

Arctic In all months temperatures are near or below freezing. Precipitation is low and mostly snow.

Sub-arctic Long cold winters. Short but quite warm summers with long hours of daylight.

Continental Typically cold winters with snow and warm summers. Low precipitation.

Cool maritime Rain occurs in all months. Rarely extremes of heat or cold. Mild winters and mild/warm summers.

Mediterranean Mild and wet winters. Hot and dry summers with storms in the autumn.

Mountain Permanent snow in mountains. Storms in summer. Warm in valley bottoms.

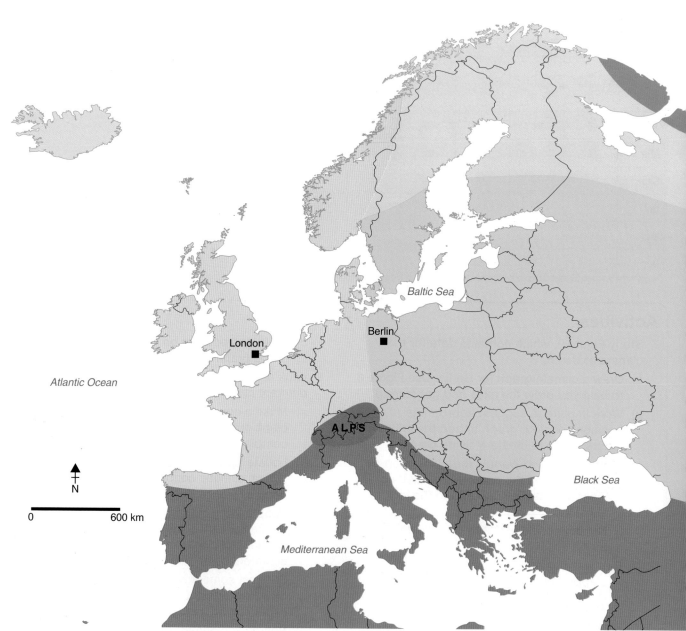

▲ **Figure 10   Climate map of Europe**

E Extreme weather: the European heatwave 2003

▲ Figure 11   The European heatwave 2003

In August 2003, Europe experienced the hottest summer for 500 years (Figure 11). Temperatures soared into the 40 °Cs for several weeks, right across southern and central Europe. In the UK, a new temperature record of 38.5 °C was recorded on 10 August near Faversham in Kent.

The heatwave caused tremendous problems across the continent, including water shortages, power failures and forest fires (Figure 13 on page 22). An estimated 17,000 people died as a result of the heatwave and many others suffered from a variety of heat-related illnesses.

Country	Deaths
France	11,000
UK	900
Spain	100
Portugal	1,300
Italy	2,000
Netherlands	1,500
Germany	300

◀ Figure 12
Deaths caused
by the heatwave
2003

Activities

9 Study Figure 11.

a) What evidence is there in the photograph to suggest that the weather is very hot?

b) Suggest an alternative caption for the photograph. Imagine you are writing the headline for the front page of a newspaper featuring the photograph.

c) Whilst the heatwave caused many serious problems (Figure 13, page 22), some groups of people benefited. Suggest how each of the following benefited from the long hot summer of 2003:

● ice-cream companies
● hotel owners
● bottled water manufacturers
● farmers producing salad vegetables and tomatoes.

d) Can you suggest any other groups of people who may have benefited from the heatwave?

10 Study Figure 12.

a) Draw a pie chart to represent the numbers in the table. To convert the numbers to degrees you need to divide each figure by the total (17,100) and then multiply by 360. Round your answers to whole degrees and check that the total is 360.

b) Use colours and a key to make the pie chart more attractive and easier to interpret.

c) Most of the deaths occurred in southern European countries. Why do you think this was so? Why do you think there weren't as many deaths in Spain?

d) The total death toll was incredibly high. Why do you think so many people died?

e) How do you think individual people should prepare for similar heatwaves in the future?

f) How do you think governments should prepare for similar heatwaves in the future?

Rivers and lakes dried up causing severe problems to aquatic wildlife and leading to problems of water supply. In the UK, hosepipe bans were introduced.

With insufficient running water, hydroelectric power stations were closed down leading to shortages of power in some areas.

Snow and ice melted in the Alps leading to rockfalls and landslides.

Many farm animals (chickens, pigs and cows) died in the extreme heat leading to increases in the price of meat in shops.

Forest fires broke out in many countries. In Portugal, 215,000 hectares of forest (about the same size as Luxembourg) were destroyed. Fires raged elsewhere in France, Spain and Italy. Many of the fires were started deliberately.

Rail tracks buckled in the heat and some road surfaces actually melted.

Many people suffered from heatburn and sunstroke.

▲ Figure 13 The effects of the European heatwave 2003

Activity

11 Study the information in Figure 13 about forest fires.

a) Why do you think there were so many forest fires?

b) Why do you think the fire services found it difficult to put out the fires?

c) Most of the fires were started deliberately. This is a criminal offence called arson. Why do you think some people chose to do such a stupid thing?

d) Forest fires do occur naturally (most commonly from lightning strikes). Can you think of any ways that fires might benefit forest ecosystems?

F Issue: How can water supply in Mallorca be managed sustainably?

Where is Mallorca?

Mallorca is the largest of the Balearic Islands in the Mediterranean Sea (off the islands of Menorca and Ibiza). Mallorca is an extremely popular tourist destination. This is because it enjoys a warm and sunny climate and has many natural attractions for tourists.

Sierra de Tramuntana

Palma in Mallorca

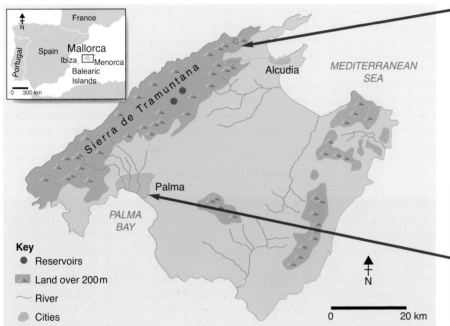

Key
- ● Reservoirs
- ▨ Land over 200 m
- ∼ River
- ◗ Cities

▲ Figure 14 Mallorca

Where does Mallorca get its water?

Mallorca receives about 400 mm of rainfall a year. This is about two thirds of the rainfall of London and less than half the rainfall of Bristol.

Most of the rain in Mallorca falls in the autumn during September and October. For much of the summer, from May through to the end of August, Mallorca has very little rainfall indeed. It is during this period that water shortages are at their most serious.

The majority of water in Mallorca comes from groundwater supplies, which are deep within the underlying rocks. The rocks in Mallorca are mostly **permeable**, which means that they allow water to pass through them. When the water gets in the rocks, it is stored in cracks and holes. Figure 15 shows a pumping station that draws water to the surface from deep underground. Some water is stored in reservoirs (see Figure 14), but these are dependent on rainfall and a large amount of water is lost by evaporation during the hot summer.

▲ Figure 15 Groundwater pumping station, Mallorca

Why has demand for water increased recently?

Since the 1950s, tourism has grown enormously (Figure 16), as has the resident population. Demand for water has soared both for drinking and for watering gardens and golf courses (Figure 17). With increasing numbers of people demanding more and more water, the natural supply has been under serious threat. With the possibility of higher temperatures in the future (due to global warming), there has been an urgent need to manage Mallorca's water resources.

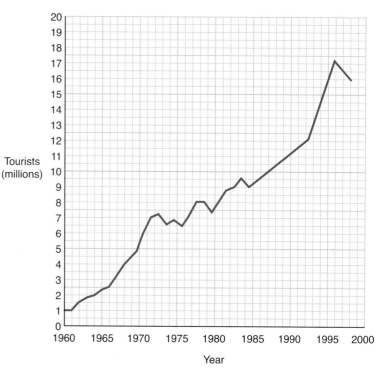

▲ Figure 16   Growth in tourists visiting Mallorca

What are the problems affecting water supply?

One of the main problems of water supply is that the demand by tourists is mostly at the coast in the south, which is a long way from the reservoirs in the mountains to the north. Transporting water is difficult and expensive.

Much of the demand for water from tourism and farming is in the summer when rainfall is at its lowest.

As demand for water has increased, some coastal groundwater supplies have become contaminated by seawater. This happens when the water levels fall below sea level. Water shortages became so severe from 1995 to 1998 that tankers transported fresh water daily from the Ebro river basin in northeast Spain to Mallorca.

▲ Figure 17   Garden with fountains in Mallorca

How have the problems been solved?

Mallorca has responded to the problem of water shortages in a number of ways:

- Increasing water supply by sinking new wells in the north of the country.

- The network of water supply pipes has been upgraded to cut down on leakages.

- Wastewater is now reused and recycled. It is reused for non-drinking water purposes, such as watering gardens and golf courses, and for irrigating crops.

- A second supply system has been introduced in Palma, carrying lower-quality reused water for non-drinking purposes. This has led to a reduction of 20 per cent in the demand for fresh water.

- In 1995, the first desalination plant was opened to treat some of the contaminated groundwater. In 1999, the Palma Bay seawater desalination plant was opened close to the capital (Figure 18). **Desalination** is a process that removes the salt from seawater. It is an extremely expensive process and can involve filtering, boiling or separating the salts by electrolysis.

The local government in Mallorca now has an effective plan to manage water sustainably for the future.

ICT ACTIVITY

Mallorca is just one place in Europe that faces problems with water supply. Access the European Commission funded 'Watertime' website at www.watertime.net/wt_cs_cit_ncr.asp . You will find a map with many locations of water supply case studies. Choose a location that interests you and find out about the water supply issues by clicking the appropriate icon. What are the problems of water supply in your chosen location and what are the options for the future?

▲ **Figure 18 Palma Bay desalination plant, Mallorca**

Activities

12 Study Figure 19.

a) How has the number of golf courses increased since 1980?

b) Describe the location of the golf courses in 2008.

c) Why do you think golfing has become so popular in Mallorca?

d) Why do golf courses in Mallorca require huge amounts of water?

e) Many golf courses are watered using recycled water. Do you think this is a good idea? Explain your answer.

13 Alcudia is a town in northeast Mallorca (Figure 14, page 23). It has suffered from serious water shortages in recent years. A local developer wants to construct two new golf courses to bring tourists to the area. A local landowner wants to expand his fruit farm on the outskirts of the town. The ferry port is being enlarged and more tourists are expected in the next few years. At present the only source of water for the town comes from an ancient groundwater supply. This is beginning to become contaminated with salt water. There are no other sources of groundwater nearby.

You are the chief water engineer and it is your job to supply water to the town in the future.

(Alternatively, you could work in a small group representing a committee.) Work through Activities a) – e) to consider your options. Then attempt Activity f) to present your report to the local government's planning committee.

a) What are the advantages and disadvantages to Alcudia of the proposed developments? Should you oppose them or try to accommodate them by increasing water supplies?

b) What is happening to the current groundwater source of water? What does this mean for the future?

c) What are the advantages and disadvantages of building a reservoir nearby? Where would you build it?

d) Do you think a desalination plant is the answer? Find out more about desalination using the internet.

e) What about recycling water? Could a campaign of water conservation be the sustainable option for the future?

f) Now write your report giving your recommendations. Include full reasoning for your preferred option. Use the internet to find maps and photographs to illustrate your report.

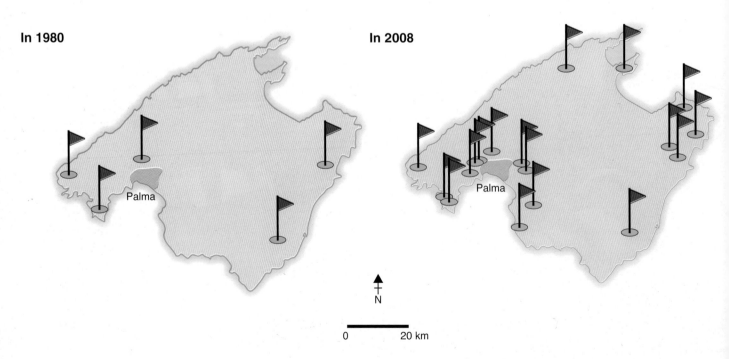

In 1980

In 2008

Palma

Palma

N

0   20 km

▲ Figure 19   Golf courses in Mallorca

Ecosystems

In this chapter you will study:

- European ecosystems known as biomes
- life in the taiga biome
- the Pyhä-Luosto National Park, Finland
- how the taiga biome is being exploited by people
- life in the Mediterranean biome
- wildfires
- protecting the Camargue from sea-level rise.

A **European ecosystems**

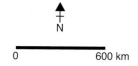

Taiga Most of the trees are coniferous (cone-bearing) and many are evergreens. They are sturdy trees capable of surviving harsh winter weather conditions. Animals include deer, wolves and bears.

Temperate deciduous forest The milder winters in these areas result in a greater variety of trees, including deciduous trees (trees that seasonally drop their leaves) such as oak and birch. The forest supports a great variety of plants, birds and animals, such as owls, foxes and deer.

Tundra This cold and dry environment supports low-growing plants and shrubs. In parts of Europe there are low-lying bogs and ponds with fish and insects. Animals found in the tundra include reindeer, rabbits and wolves.

Key

- Tundra
- Mountain (tundra)
- Taiga forest (mostly conifero
- Temperate deciduous forest
- Grassland
- Chaparral (Mediterranean)
- ★ Pyhä-Luosto National Park

Grassland Grasslands are semi-arid areas where there is not enough rainfall to support trees. These are excellent natural environments for grazing animals. Typical animals include antelope, ground squirrels and rabbits.

Mountain (tundra) Similar conditions found in northern tundra regions. The climate high up in the mountains is very similar to the sub-Arctic climate.

Chaparral (Mediterranean) In this hot and dry biome plants and animals have to cope with drought in the summer. Much of the land is bare and rocky with the occasional tree or shrub. Olives and citrus fruit (e.g. oranges and lemons) are well adapted to this environment. Lizards are common and can often be seen scuttling up walls.

▲ **Figure 1 European biomes**

An **ecosystem** describes the interactions between living organisms and the environment where they live.

In Book 1 we introduced the idea of ecosystems by studying hedgerows. Hedges are examples of small-scale ecosystems. Others include ponds and woodlands. Organisms such as plants, birds and animals form part of these ecosystems, as does the soil and the climate.

Ecosystems can also be identified at a larger global scale. These are called **biomes**. Well-known biomes include tropical rainforests and deserts. Figure 1 shows the biomes that exist in Europe.

Biomes extend over large areas and usually reflect particular climatic characteristics. They describe the natural vegetation that would be expected in a particular area. Of course, in much of Europe people have altered the natural environment considerably, which explains why in many places the natural vegetation does not exist. For example, we know that the UK is not entirely covered by deciduous forest!

Compare Figure 1 here with the map in Figure 10 from Chapter 1, on page 20, of European climate zones. Notice that the patterns on the two maps are very similar. This is because temperature and precipitation often determine the type of vegetation that grows in an area. Trees grow well in wetter climates, whereas grasslands prefer moderate amounts of rain. Around the Mediterranean Sea, the hot and dry summers account for the presence of citrus fruit trees, which are capable of surviving these harsh conditions.

Activities

1 Study Figure 1 and Figure 10 from Chapter 1 (page 20).
   a) Locate the UK. What type of biome covers the UK?
   b) What type of climate corresponds with the UK's biome?
   c) What type of biome corresponds with a Mediterranean climate?
   d) Describe the location of the taiga forest biome in Europe.
   e) Does the taiga forest biome correspond with the Arctic or the sub-Arctic climate?
   f) Where would you go in Europe to study the tundra biome?

2 Study Figure 1 and 2. Figure 2, is a section across Europe as shown on Figure 1.
   a) Make a careful copy of Figure 2, leaving plenty of space above the diagram for labels.
   b) Locate and label the Alps.
   c) Locate and label the Mediterranean Sea, the Baltic Sea and the Arctic Ocean.
   d) Now use Figure 1 to correctly locate and label the different biomes.
   e) Add some colour to your diagram to make it attractive.
   f) If you wish, you could use the internet to find some photographs to illustrate your diagram.

A   B

▲ **Figure 2   Cross-section (along A–B on map) showing European biomes**

B Life in the taiga biome

The taiga forest biome stretches in a broad band across from Scandinavia (Figure 1). It continues right around the world, through Russia and across much of Alaska and Canada (Figure 3). It is in fact the world's largest biome.

Despite the long hard winters and the short summers, the taiga is home to a wide variety of plants and animals that have adapted to cope with the harsh conditions. Birds with strong beaks hunt for insects and pine nuts (Figure 4), and squirrels scamper in the branches. Insects abound in the summer (taiga forests are well known for the swarms of mosquitoes!). Many birds leave when the cold weather comes, and some animals hibernate.

Taiga forests are beautiful, magical places that come alive in the spring and summer. Open forest glades support wild flowers, and in the autumn the forests are bursting with berries. They are very popular with people who enjoy the outdoor life.

Activity

3 Study Figure 5, which is a sketch of a typical evergreen tree found in the taiga.

a) Make a large careful copy of the sketch.

b) Add the following labels to describe some of the characteristics of the tree that help it to survive in the harsh climate.

- Needle-like leaves help to prevent water loss during dry periods
- Needles help the tree to shed snow
- Branches droop so are less likely to break after heavy snowfall
- Trees are evergreen and do not lose their leaves in the winter. In the spring energy does not have to be used to grow new leaves
- Thick bark resists damage to the tree during wildfires.

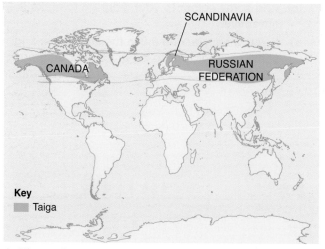

▲ Figure 3 The global distribution of taiga

▲ Figure 4 Woodpecker on coniferous tree in a taiga forest in Finland.

Activities

4 Study Figure 6, which shows the arctic hare. This is a common animal in the taiga forest.

a) Why do you think arctic hares have thick fur?

b) Why do you think arctic hares have large furry feet?

c) In summer arctic hares have grey fur but in winter their fur turns white. Why?

d) Arctic hares are very fast and can quickly change direction. They can even swim (hare paddle of course!). How do you think these adaptations help them to survive?

e) Arctic hares normally eat grass but they can survive by eating bark and buds. How does this help them to survive through the year?

f) If the climate becomes warmer and there is less snow, how do you think arctic hares might adapt in the future? Give reasons for your answer.

5 Imagine that you are a biologist researching the behaviour of arctic hares in the winter. You are to spend a week in a log cabin in northern Finland. There is no electricity or running water but there is a nice hot sauna!

a) What aspect of the arctic hare's behaviour would you be interested in studying and why?

b) Work with a friend to make a list of the clothes and provisions that you would need to take with you to survive the week.

c) What do you think you would enjoy most about the experience?

d) What do you think you would miss most whilst stuck in a log cabin for a week?

e) Why do you think saunas are so popular in Scandinavian countries?

ICT ACTIVITY

There are several very good websites that outline the adaptations of animals to living in the taiga forest biome. Have a look at the following sites and choose one animal that interests you. Find a photograph of the animal and add labels to describe its adaptations to the conditions experienced in the taiga forest. The results of your research could form a class display on a wall.

www.blueplanetbiomes.org/taiga.htm

www.mbgnet.net/sets/taiga/index.htm

▲ Figure 5 Taiga tree

▲ Figure 6 Arctic hare

C Map Study: Pyhä-Luosto National Park, Finland

Pyhä-Luosto National Park is located in Northern Finland (Figure 1, page 28) in the Scandinavian region called Lapland. It was established in 2005, and is an area of great beauty (Figure 7). The landscape is relatively hilly, although most of Finland is very flat. The hills are the remains of very ancient mountains formed 2,000 million years ago. Much of the land is covered in pine forests, some of which are over 200 years old.

Figure 8 is a map extract of the area showing the National Park. A thick green line shows the boundary of the National Park. Some of the symbols are explained in the key. Reindeer are herded in the area. The word Poroerotus on the map locates a wooden corral where the reindeer are collected.

Activities

6 Locate the line of hills to the west of the main road. What is the highest point in these hills?
   a) Use the key to describe the natural vegetation of the area.
   b) There are several green dotted lines in the National Park. What are they?
   c) What are the attractions for tourists in Luosto?
   d) You are visiting the area shown on the map and want to go camping. Where would you choose to go?
   e) What is the evidence that this area has snow in the winter?
   f) There are lots of lakes on the map extract. Why might they be popular with tourists?

7 You have been asked to produce a single-sided leaflet advertising Pyhä-Luosto National Park to UK tourists. The leaflet will be made available in Scandinavian tourist information offices. Make use of Figure 7 and your answers to the questions in question 6. You will also find some interesting tourism information on the internet. Try the following websites to get you started.
   www.luontoon.fi/page.asp?Section=5248
   www.laplandluosto.fi/winter/summeractivities/attractions/pyhaluostonationalpark

▲ Figure 7 Pyhä-Luosto National Park

Figure 8 Map extract of Pyhä-Luosto National Park ▶

0 _____ 10km
            scale

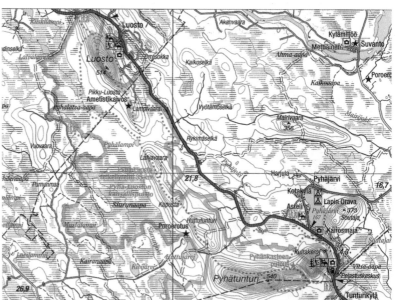

Key

8,5	Distance (km)
24h	Service station 24hrs, filling station
	Hotel or motel, other accommodation
	Youth hostel, holiday village, camping site
⚓ ★ ★	Monument, other special point of interest, natural formation
	Entertainment site, sports centre, winter sports
	Nature information centre, access to nature park or trail, bird-watching tower
=====	Nature trail
....	Snowmobile track
Opisto	Notable building, view tower, mast
	Field, wooded marsh, open marsh

D Exploiting the taiga

People have exploited the taiga forests in Europe for hundreds of years. Indeed, much of the original forest has now been cut down and what we see today is secondary growth. Unfortunately, the secondary forests contain far fewer plant and animal species than the original forest. Today the forests are managed in a **sustainable** way, with more trees being planted than being felled.

Most forests are exploited for timber, with the wood being used for a variety of purposes. Some is used in the building trade and for making furniture (Figure 9). Some of the wood is pulped and then used to make paper (Figure 10). Increasingly, recycled paper is added to the natural pulp in papermaking. Today, about 50 per cent of the fibres used to make paper comes from recycled paper and this is expected to increase in the future.

Scandinavian forests are extremely popular with local people who enjoy trekking and camping. The forests are well suited to the growth of berries, such as blueberries and bilberries. Some 60 per cent of Finnish households engage in berry picking, harvesting an average of 25 kg per household. Local food industries benefit from the berries by making jams and other products.

▲ **Figure 9 Furniture making**

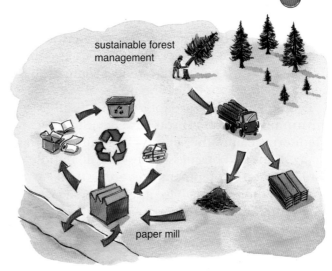

sustainable forest management

paper mill

▲ **Figure 10   Paper mill and forest management**

Activity

8 Study Figure 10.

   a) Make a copy of the diagram in Figure 10.

   b) Add the following labels in the correct places.
   - Trees cut down and loaded onto lorries
   - Wood chips and thinnings to make paper
   - Timber products (e.g. planks)
   - Used paper and board recycled

   c) Why do you think the paper mill is located next to a river?

   d) Is there any evidence on Figure 10 that new trees are being planted to replace those being cut down?

   e) Trees absorb carbon dioxide. Can you suggest why this is a good thing?

   f) More and more recycled paper is being used to make paper. Do think this is a good thing? Explain your answer.

 ICT ACTIVITY

Find out more about the types of berry found in Finland's forests by accessing the website www.dlc.fi/~marianna/gourmet/i_berry.htm.
- Which berries are eaten fresh?
- Which berry turns a bright yellow-orange when it is ripe to eat?
- Which berries are used to make jam?
- You have bought a jar of Lingonberry jam at a sale. How could you make use of jam apart from spreading it on toast?
- Select a type of berry of your choice. Draw a picture to show what it looks like. Describe where and when it can be picked. How can it be used in the kitchen?

E   Life in the Mediterranean biome

Plants and animals living in the Mediterranean biome have to cope with a hot dry summer and this often brings with it the problem of drought. As a result, the plants and animals have adapted to cope with these difficult conditions.

Most of the Mediterranean biome is made up of low-growing shrubs and woodlands (Figure 11). This is called **maquis**. Many animals inhabit the Mediterranean biome, such as wild goats, rabbits, wild boar, sheep and lynx. Insects, bees and lizards are also common. One such lizard is the Mediterranean gecko (Figure 12). In common with many Mediterranean animals it is small and **nocturnal**. Small animals are well adapted to lose heat and they can also burrow more easily when seeking shelter from the sun. They feed during the night when it is cooler. The animals are often light in colour. This helps to reflect some of the sunshine and stops it heating up.

People have had to adapt to the natural environment as well. If you have been to the Mediterranean in the summer you will know that many shops and businesses close in the middle of the day when the temperatures are at their highest. Some people have a rest called a **siesta**. The shops and offices reopen and then stay open into the evening when it is cooler.

The Mediterranean biome provides grazing for farm animals, such as sheep and cattle. Land is also used for growing grapes for making wine. **Citrus fruit** (such as oranges, grapefruit and lemons) grow well and olive groves (for olive oil) are particularly widespread in parts of Italy (Figure 13).

▲ Figure 12   Mediterranean gecko

▲ Figure 13   Olive tree

Activity

9   In this activity you are going to design a new tree or a new lizard that is well adapted to Mediterranean conditions.

a)   Re-read the information in the text describing how trees and lizards have adapted to the hot and dry summers.

b)   Draw your own made-up tree or lizard that has the adaptations described in the text. Use colours if you wish.

c)   Add labels to show the main features of your tree or lizard. Describe how each feature will help your tree or lizard to survive.

d)   Think of a name for your new tree or lizard and use it in writing a title.

Some trees have thick bark to make them fire-resistant

Trees have long 'taproots' to search for water deep underground

Plants often have small waxy leaves to reduce water loss

Shrubs often have a network of fine roots spreading out horizontally close to the surface to obtain water as it falls on the ground

▲ Figure 11   Mediterranean shrublands (maquis)

F Wildfires in the Mediterranean biome

One of the biggest threats to any land ecosystem is fire. Wildfires (often started by lightning strikes) are common throughout the world, but particularly in regions that experience long dry periods.

The Mediterranean biome often has wildfires. In most summers fires break out in parts of Portugal, Spain, France, Italy and Greece. The vegetation is often extremely dry during the long hot summer and it ignites easily. Some plants actually contain flammable oils!

In 2007, in Greece, there were a number of fires close to the capital, Athens (Figure 14). In total some 200,000 hectares (1 hectare is about the size of a football field) including trees and olive groves were destroyed. Sixty-seven people died – many incinerated in their cars when trying to escape. Thousands of houses were destroyed.

The wildfires in Greece resulted from a long period of severe drought and extremely high temperatures up to 45 °C. The vegetation was dry and burned easily. Winds fanned the flames and caused the fires to spread out of control.

Fires pose a number of threats to natural ecosystems and to the people who live in the areas. However, they also bring benefits by getting rid of dead wood, clearing an area of pests and helping certain plants to germinate. Figure 15 describes some of the advantages and disadvantages of wildfires.

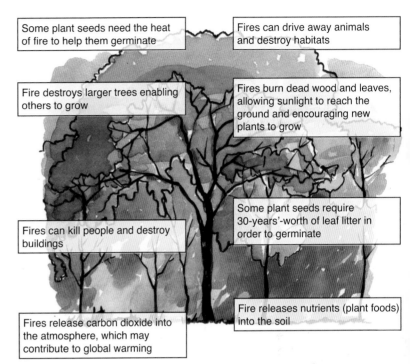

- Some plant seeds need the heat of fire to help them germinate
- Fires can drive away animals and destroy habitats
- Fire destroys larger trees enabling others to grow
- Fires burn dead wood and leaves, allowing sunlight to reach the ground and encouraging new plants to grow
- Fires can kill people and destroy buildings
- Some plant seeds require 30-years'-worth of leaf litter in order to germinate
- Fires release carbon dioxide into the atmosphere, which may contribute to global warming
- Fire releases nutrients (plant foods) into the soil

▲ **Figure 15 Advantages and disadvantages of fires**

Activities

10 Study Figure 14.
  - **a)** What types of vegetation is the fire affecting?
  - **b)** Is there any evidence that the fire is being fanned by winds?
  - **c)** What do you think is the effect of the fire on people and human activities?
  - **d)** The photograph in Figure 14 has been chosen to appear in a newspaper. Write a few sentences to accompany the photograph. What is happening in the photograph? Describe what it feels like to be close to where the photograph was taken.

11 Study Figure 15.
  - **a)** Use Figure 15 to draw up a table to list the advantages and disadvantages of fires.
  - **b)** Identify a disadvantage that involves people.
  - **c)** Which disadvantage affecting the natural ecosystem do you think is most serious and why?
  - **d)** What is the effect of wildfires on the climate?
  - **e)** Wildfires tend to be left to burn themselves out naturally, unless they threaten people or property. Do you think this is a good idea? Explain your answer with reference to the advantages listed in Figure 15.

▼ **Figure 14 A forest fire in Corinth, Greece, 2007**

Some Mediterranean plants and trees have developed adaptations to make them resistant to wildfires. Many sprout from underground roots, and some have seeds that require the heat of a fire to crack them and allow germination.

One interesting tree that is well adapted to coping with fire is the cork oak (Figure 16). This is an amazing type of oak tree that has a very thick bark. It is this bark that protects the tree from fire. The cork oak has an important commercial function. Its bark is used to make the corks that act as stoppers in wine bottles. Once removed (every 9 or 10 years), new bark will grow to protect the tree from future fires and to provide much-needed corks for the wine industry.

▲ Figure 16 Cork oak tree

Activity

12 Study Figure 16.

a) What evidence is there that bark is being stripped off the cork oak tree?

b) Why do you think cork bark is first removed towards the base of the tree?

c) How do you think the men will get to the higher branches?

d) Why do you think there is not much other vegetation in the photograph?

e) Can you suggest an advantage that the cork oak tree brings to the ecosystem in the photograph?

f) Can you think of a disadvantage?

g) If a fire were to break out now, why would the tree be particularly vulnerable?

ICT ACTIVITY

Find out more about the cork industry. Access Jelinek Cork's website at www.jelinek.com/about_cork.htm. Based in Canada, Jelinek Cork has been operating across the world for 150 years.

- Click 'About Cork'.
  a) Where in the world do cork oak trees grow?
  b) Apart from fires, how are cork oaks well suited to living in the Mediterranean biome?

c) Describe how cork is harvested from the tree. Include a labelled photograph.

d) What are the characteristics of cork that make it a useful product?

e) list some of the products

- Click 'Product Information'. Design a collage spread to show the huge range of products that can be made using cork. Include simple sketches or photographs to make your collage attractive. You could even decorate your collage with actual cork products, such as wine bottle corks!

G Issue: How can the Camargue be protected from rising sea levels?

Where is the Camargue?

The River Rhône is one of Europe's major rivers. From its source high up in the Swiss Alps, it flows west, and then south (through France) to reach the Mediterranean Sea near Marseilles. At Arles, the Rhône splits into two channels, the Grand Rhône to the east, and the Petit Rhône to the west (Figure 17). In between the two channels is found one of Europe's most valuable ecological wetlands – the Camargue.

What makes the Camargue special?

The triangle of land that forms the Camargue is the Rhône delta (Figure 17). It is an area of marshland and lagoons formed by river deposition over thousands of years. The Camargue is home to an astonishing variety of plants and animals. It is particularly well known for its flocks of flamingos, its white horses and its bulls (Figure 18).

The Camargue is a regional nature park and its varied activities are carefully managed to prevent conflict and avoid harm to the environment. There are many demands on the Camargue, including tourism, farming and the production of salt. Only 60,000 people live in the Camargue, 85 per cent of whom live in the town of Arles.

What is the threat to the Camargue?

The main threat facing the Camargue in the future is a rise in sea levels, due to global warming. Scientists believe that sea levels in the area could rise by 1m by 2050 and by as much as 5m by 2100. Such a rise in sea level could have a massive impact on the geography of the area.

▲ Figure 18 The Camargue

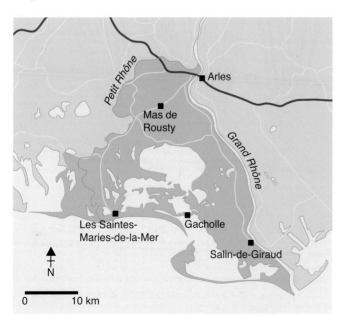

▲ Figure 17 Location map of the Camargue

▲ Figure 19 Sand dune restoration

How can the Camargue be protected?

In 1859, a sea dyke (low wall) was constructed at the outer edge of the delta to protect it from the sea. Embankments have been built since then to strengthen the defences.

Recently, attention has turned to the sand dunes that line much of the coastline. If these natural sea defences could be built up further, they would help to protect the Camargue from sea-level rise.

The sand dunes at Piemanson near Salin-de-Giraud (Figure 17, page 37) have been damaged by trampling and by erosion from the sea. The park authority has built sand traps (Figure 19) arranged to hold back the sand. Marram grass has been planted to trap the sand and encourage the dunes to increase in height. Marram grass is well adapted to living in sand dunes and it occurs naturally in these hostile environments (Figure 20).

▲ Figure 20   Marram grass

Activity

13 A stretch of beach near Les Saintes-Maries-de-la-Mer (Figure 19) is backed by sand dunes. Over the years, tourists have damaged the plants and the sand is beginning to be blown away by the wind. The park authority is unsure whether to build a sea wall to protect the area from sea-level rise, or opt for a more ecological solution involving planting marram grass. You have been asked to put together an argument in favour of planting marram grass. Use the internet to find out more about the Camargue, sea walls, and the advantages of planting marram grass when restoring sand dunes.

a) You can present your argument in any format, e.g. hand-written sheet, ICT-produced brochure using Word or Publisher, or a Powerpoint presentation.

b) Include text, photographs and diagrams to outline the reasons why an ecological approach involving the use of marram grass may be a better option than building a sea wall.

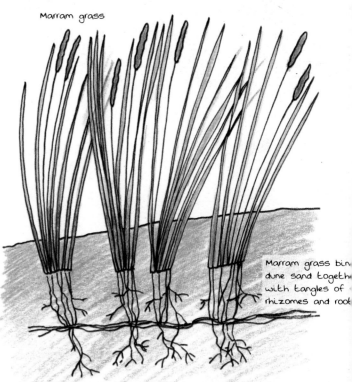

Marram grass grows vigorously (up to 1 metre a year) when buried by sand. It has long tangled roots that help to anchor it to the sand. They are capable of searching for water deep below the ground surface. It is able to cope with saltiness in the air. Its thick and waxy leaves help to reduce water loss and help the plant to survive hot and dry conditions. The dense leaves help to trap blowing sand.

▲ Figure 21   Marram grass

Mountains

In this chapter you will study:

- mountains in Europe
- ice in the mountains
- the Mer de Glace glacier
- avalanche!
- volcanic mountains: Mount Etna
- problems facing ski resorts due to global warming, and possible solutions.

A European mountains

Key

▲ Mountains (all heights in metres above sea level)

Colour key from atlas showing heights

| 0-200m | 200-400m | 400-1000m | 1000-2000m | 2000-4000m | 4000-5000m | 5000 + m |

▲ **Figure 1 Mountain ranges**

Europe has a huge variety of physical landscapes (see Figure 12, page 9). In this chapter we will be looking at some of the most dramatic landscapes: the mountains. We shall be finding out what opportunities and challenges they present to people who live and work in these extreme environments.

Europe has several mountain ranges (Figure 1). The highest mountain range is the Alps, which stretches from France through Italy and Switzerland and into Austria. There are some well-known mountain peaks in the Alps, such as the Matterhorn (Figure 2) and Europe's highest mountain at 4808m above sea level, Mont Blanc.

Elsewhere in Europe, the Pyrenees mark the border between France and Spain. In the east, the Carpathian Mountains stretch through Slovakia, Ukraine and Romania.

Living in the mountains can be difficult and demanding, as Figure 3 illustrates.

▲ **Figure 2   Matterhorn**

Extreme weather: heavy rain and winter snow makes life difficult

Steep slopes are difficult for farming, building and transport

Tunnels through the mountains

Landslides and avalanches are common

Expensive bridges to carry roads over wide rivers; road networks are limited in the mountains

Rivers often flood, especially when snow melts in the spring

▲ Figure 3   **Problems facing people living in the mountains**

Activities

1 Study Figure 1 on page 40 and the atlas map in the inside back cover.

a) Name and give the height of the highest peak in the Pyrenees.

b) The Matterhorn in the Alps is 330m lower than Mont Blanc. True or false?

c) Mount Olympus is the highest mountain in which country?

d) What is the name and the height of the highest mountain in Italy?

e) In which mountain range is Puy de Sancy the highest peak?

f) What is the name and height of the highest mountain in Scandinavia?

g) In which country is it located?

h) The Pennines and the Apennines are two upland areas in Europe. In which countries are they located?

2 Study the map extract in Figure 4 below. It shows a small part of the Alps close to Mont Blanc in northern Italy. Look closely at the route of the main road, the S 26d. The engineers who built the road had to deal with several problems. The labels **A**, **B** and **C** show the location of three of them.

a) What is happening to the road at **A**? Why do you think the road engineers decided to do this with the road?

b) What is the name of the river that passes beneath the bridge at **B**?

c) Why did engineers have to build such a wide bridge at **B**?

d) At **C** the road doubles back sharply. This is called a 'hairpin bend'. Why do you think the engineers had to follow this route rather than just building a straight section of road?

e) Apart from problems with road transport, what other difficulties do you think face people living in Entrèves? Use Figure 3 on page 41 to give you some ideas. Try to refer to specific details on the map to support your answer.

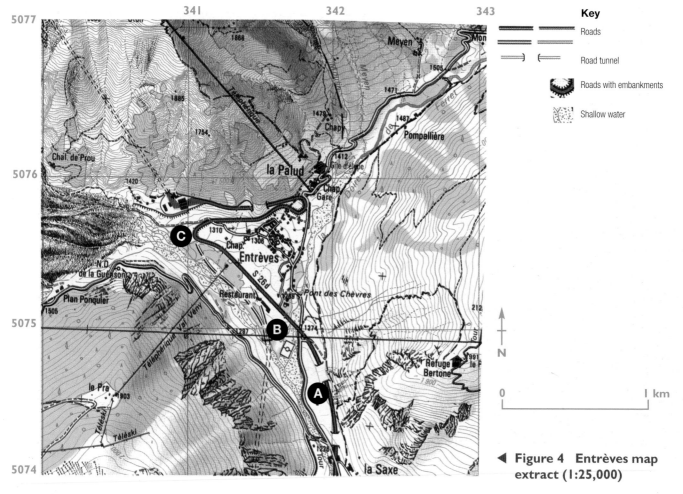

Key

Roads

Road tunnel

Roads with embankments

Shallow water

◀ Figure 4 Entrèves map extract (1:25,000)

B Ice in the mountains

Look at Figure 5. It is a satellite photograph of part of Western Europe. Look closely at the mountains to see the patches of white. These are areas of snow and ice. If we zoom in on the Alps (Figure 6), it is now possible to see individual rivers of ice called glaciers spreading down the mountainsides like the fingers of a giant hand. These glaciers have helped shape the Alps, forming the spectacular mountain landscape than many of us enjoy.

Activity

3 Study Figures 5 and 6.

a) Figure 5 shows part of the Alps. Which country is shown in the satellite image?

b) High up in the mountains near Mont Blanc (Figure 6), the surface is very white. Can you explain why?

c) Locate the Mer de Glace on Figure 6. It is mostly grey in colour. What does this suggest to you about the surface of the glacier?

d) Is the Mer de Glace flowing north or south? Explain your answer.

▲ Figure 5 Satellite image of Western Europe

▲ Figure 6 Alps near Chamonix, showing Chamonix and the Mont Blanc range with the Mer de Glace

43

C The Mer de Glace

Look at Figure 7. It shows part of the Mer de Glace glacier near Chamonix in France (Figure 6, page 43). The Mer de Glace (what do you think this name means in English?) is one of the largest and most spectacular glaciers in the Mont Blanc range. It is about 7 km long and up to 200 m deep. It moves downhill at about 90 m a year, that is 1 cm per hour!

Notice the brilliant white snowfields in the far distance. It is here, high up amongst the jagged peaks, that the Mer de Glace has its source. Over many years, the layers of snow are compressed by freshly falling snow and they slowly turn to solid ice. As the ice mass grows in size, it starts to move downhill (due to the effect of gravity). Spilling out of its mountain hollows, it begins to flow along old river valleys (such as the one you can see in Figure 7).

Activity

4 Study Figure 7.
   a) The mountain peaks in the distance are locally known as 'aiguilles'. This is the French word meaning 'needle'. Do you think this is a good name to use for the mountain peaks? Explain your answer.
   b) Make a list of five words to describe the landscape in the photograph.
   c) Imagine that you have bought a postcard of this view to send home. Use the five words to help you write a few sentences describing the scene.
   d) Tourists have visited the Mer de Glace and looked at this view for over 100 years. Why do you think it is so popular with tourists?

▲ **Figure 7** Mer de Glace

People often comment, when seeing the Mer de Glace for the first time, how dirty it looks! What do you think? Take a close-up view of the surface of the glacier by looking at Figure 8. Many of these rocks have tumbled down the valley sides, having been loosened by freeze-thaw weathering. In the past, some of the rocks were buried beneath layers of snow, and were transported down the valley by the moving ice. When the upper layers of ice melted, they then became visible on the surface. The glacier acts like a massive conveyor belt transporting ice and rocks down the valley.

Can you also see the big crack in the ice in Figure 8? This is a **crevasse**. Crevasses are common on the Mer de Glace and they form when the surface ice is stretched as the glacier passes over a step in the valley floor. Crevasses are very dangerous, especially in the winter when they are hidden by a thin layer of snow. Notice, too, the blue colour of the ice – this is typical of glacier ice and is caused by sunlight being refracted as it passes through the ice.

▲ **Figure 8 Mer de Glace close-up**

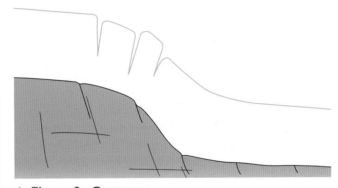

▲ **Figure 9 Crevasses**

Activities

5 Study Figure 8.

  a) Are the rocks all the same size or are they very mixed?

  b) Are the rocks smooth and rounded or jagged and angular?

  c) Draw a sketch of one of the larger rocks to show its shape. Add a label to your sketch.

  d) Suggest reasons why it is very dangerous to walk on the ice.

  e) Design a poster to be displayed at the edge of the glacier to warn people not to walk on the ice. You must keep the poster simple and very visual because tourists come from many countries and speak different languages. Think up a clear message and be very direct. You can use ICT if you wish.

6 Study Figure 9. It is a diagram showing the formation of a crevasse.

  a) Make a copy of the diagram in Figure 9.

  b) Add a label to identify a crevasse.

  c) Add a few rocks and boulders to the surface of the ice, using Figure 8 to help you.

  d) What do you notice about the slope of the valley floor beneath where the crevasses have formed?

  e) Why do you think the crevasses have formed at this location on the glacier?

ICT ACTIVITY

Every year, an unfortunate few walkers and mountaineers fall down a crevasse.

Use the internet to find out how people can be rescued from crevasses. Present your information in the form of a 'What to do …' rescue guide. It should take up no more than a single side.

Here are some useful sites, although a search 'crevasse rescue' will reveal many others.

NOVA at Public Service Broadcasting at www.pbs.org/wgbh/nova/denali/extremes/survcrev.html

Planet Fear at www.planetfear.co.uk/articles/Crevasse_Rescue_Technique_370.html

Wikipedia at http://en.wikipedia.org/wiki/Crevasse_rescue

The Mer de Glace is very popular with tourists who often visit from nearby Chamonix. Each year a short tunnel (known as the 'ice grotto') is drilled into the ice to let visitors explore the insides of a glacier (Figure 10). It is amazing to see and touch the clear blue ice. You can see bubbles and even pieces of rock trapped in the ice. All around water drips from the roof onto the floor. Sometimes you can even hear the ice cracking and creaking. Looking down the glacier from the tunnel entrance, previous years' tunnels can be seen in the distance, clearly showing how the glacier is ever so slowly moving downhill.

▲ Figure 10 Ice grotto in Mer de Glace

Activity

7 Study Figure 10.
   a) Why do you think there is a carpet on the floor of the ice grotto?
   b) How do you think it would feel to touch the ice as the people are doing in the photo?
   c) Why does a new grotto have to be drilled each year?
   d) There is a notice outside the entrance giving advice to people entering the grotto. What advice do you think is given to people and why?
   e) If you visited the Mer de Glace would you want to go inside the ice grotto? Give reasons for your answer.

D Avalanche!

What is an avalanche?

Peggy Harris, a skier who was caught in an avalanche in the Alps, wrote the following account.

"I heard a big explosion that stopped me in my tracks. At first I thought it was a plane, but then I looked up and saw a huge cloud of snow coming towards me. I realised it was an avalanche. My legs were suddenly blown out from beneath me. It was like being in a washing machine. The pummelling you take is unreal. Finally, I came to a stop. I was able to move my hands, but I was trapped. Buried alive. Then I blacked out."

Peggy survived her ordeal because she was wearing a transmitter that pinpointed her position in the snow. This enabled rescuers to find her and dig her out. She was very lucky.

Avalanches are quite common in the Alps and they have claimed the lives of many people over the years. Figure 11 shows a typical avalanche. Notice the cloud of broken snow and ice in the air. What you can't see is the surge of ice and rocks close to the ground that is powerful enough to strip trees from a hillside or destroy a house. Several factors trigger avalanches (Figure 12), such as heavy snowfall, a sudden rise in temperature or people skiing '**off piste**' on freshly fallen snow.

Activities

8 Study Figure 12.

   a) How can people trigger an avalanche?

   b) Why do you think heavy snowfall increases the likelihood of an avalanche?

   c) Why do earthquakes often trigger avalanches?

   d) Scientists sometimes let off explosions to deliberately start an avalanche. Why do you think they do this?

9 Study Figure 13. It shows the typical features of an avalanche.

   a) Make a careful copy of Figure 13.

   b) Write the following labels instead of the empty boxes:

   ● backwall of the avalanche

   ● avalanche track

   ● trees that restrict the extent of the avalanche

   ● mountain peak

   ● avalanche run-out (where the avalanche stops)

   ● skiers

   c) What do you notice about the track of the avalanche and the position of the trees on the mountainside?

   d) How could this observation help to prevent future avalanches in this area?

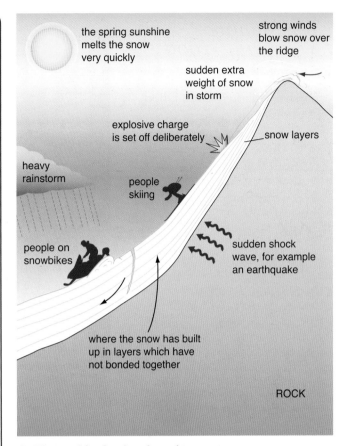

▲ Figure 12 Avalanche triggers

▼ Figure 11 Snow avalanche in the Alps

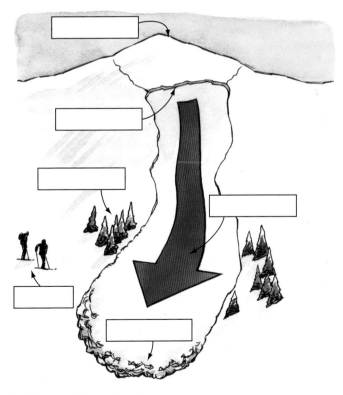

▲ Figure 13 Avalanche

The Montroc Avalanche, 1999

On 9 February 1999, a massive avalanche that killed 12 people and destroyed many houses hit the village of Montroc near Chamonix. At about 2.40pm, a giant slab of snow 1.5 m thick and covering an area of 30 hectares (1 hectare = a football field) tore down the mountainside at a speed of 100 km/hour. It crashed into the village burying the houses in up to 5 m of snow. Rescuers struggled to get to the village, as roads had been blocked by recent heavy snowfalls. The emergency services arrived to be greeted by a scene of utter devastation.

▲ Figure 14   The effects of the Montroc avalanche

Activity

10 Study the photo in Figure 14 and the details about the Montroc avalanche.

   a) The main trigger of the avalanche was very heavy snowfall. What is the evidence of heavy snowfall in Figure 14?

   b) Describe the effects of the avalanche on the wooden buildings in the photograph?

   c) Many of the houses destroyed by the avalanche had been built recently. Why do you think new houses have been built in the Alps in recent years?

   d) How do these new housing developments increase the risk of avalanches?

ICT ACTIVITY

The purpose of this activity is to put together a short Powerpoint presentation (of no more than 12 slides) to show the devastating effects of avalanches. The presentation is to be shown to skiers at an Alpine resort to encourage them to be watchful and not to ski off-piste. You can add some background music to your presentation if you wish.

Use the internet to search for some powerful and dramatic photographs.

National Geographic
http://science.nationalgeographic.com/science/photos/avalanche-general.html

Utah Avalanche Center
www.avalanche.org/~uac/obphotos/observer.html

Avalanche.org www.avalanche.org/picturepage.htm

How can the threat from avalanches be reduced?

Most people killed by an avalanche trigger it themselves. Therefore, to reduce the numbers of deaths, people need to be more aware of the dangers. They need to pay attention to warnings of avalanches given by scientists. They should not ski or snowboard off-piste (away from the managed ski tracks). They should carry small shovels and transmitters to help people find them.

Look at Figure 15. It describes some ways of reducing the impact of avalanches on human activities.

Activity

11 Study Figure 16. It shows the location of a small skiing village in the Alps. You have been asked to suggest ways of reducing the avalanche risk
  a) Make a large copy of Figure 16.
  b) Study the avalanche protection measures in Figure 15. Select and draw onto your diagram some ways of protecting the village from future avalanches.
  c) Write a few sentences giving reasons for the protection measures that you have selected.

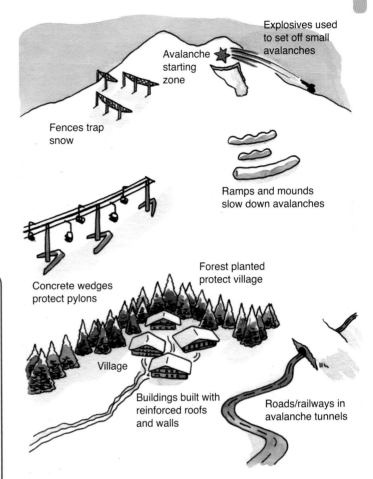

▲ **Figure 15 Ways of reducing avalanches**

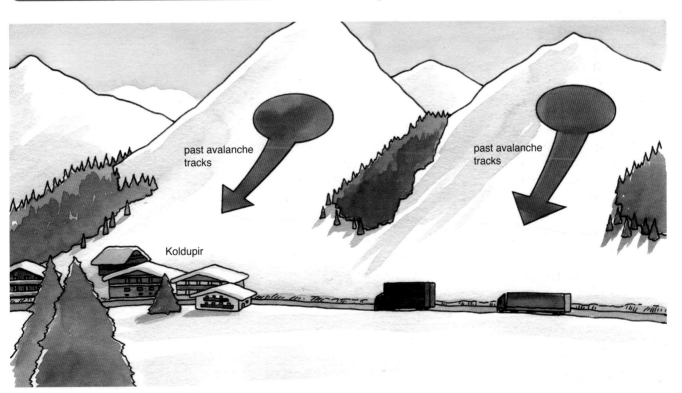

▲ **Figure 16 Koldupir**

E Volcano: Mount Etna

Some of Europe's mountains are active volcanoes. You have probably heard of Mount Vesuvius in Italy and how in AD79 it destroyed the Roman town of Pompeii.

Europe's largest active volcano is Mount Etna on the Italian island of Sicily (Figure 17). Mount Etna erupts regularly, sending torrents of lava down the mountainside. If you look closely on the map extract in Figure 18 you can see the lava fields. Notice how roads and villages have been affected by lava flows in the past.

Mount Etna last erupted in September 2007 at about 8.00pm. The eruption sent a fountain of lava some 400 m into the air, lighting up the night sky. Strong winds blew ash and smoke into nearby villages and towns. The local airport at Catania had to be closed for some time.

▲ Figure 17 Mount Etna

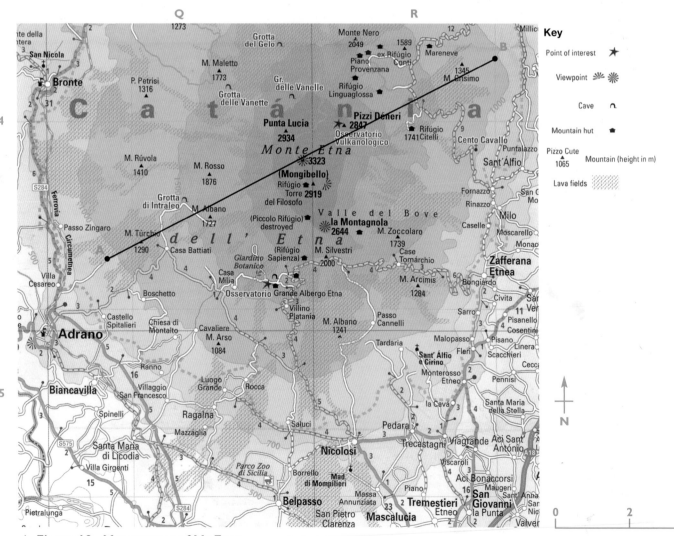

▲ Figure 18 Map extract of Mt Etna

Activities

12 Study Figure 17.

   a) Draw a sketch of the volcano from the photograph to show its shape.

   b) Add the following labels to your sketch:
- steep mountain sides
- volcano summit (top)
- landslip at the volcano's summit
- path.

13 Study the map extract in Figure 18. Notice that there are several peaks on Mount Etna. This is because eruptions have occurred in many different places on the mountain.

   a) What is the highest point on Mount Etna in metres above sea level?

   b) What symbol is used to show the location of the Osservatorio Vulkanologico (volcano observatory)?

   c) Why do you think there is a volcanic observatory on Mount Etna?

   d) Notice that there are lots of 'rifugios' on the mountain. What do you think they are?

   e) Eruptions in the past have left many caves or 'grotta' on the mountain's slopes. What is the name of the 'grotta' to the southwest of the highest peak on Mount Etna?

   f) Locate the town of Nicolosi to the south of the volcano. What is the straight-line distance from the highest peak on Mount Etna to Nicolosi?

   g) What is the evidence that Nicolosi may not be safe from future eruptions?

   h) What do you notice about the route of the main roads shown orange?

   i) Can you think of two reasons why no major roads cross the mountain? Follow the route of some of the yellow roads to give you a clue.

14 Study Figure 18. In this activity you are going to draw a cross section of the volcano.

   a) Locate the line of section on the map in Figure 18 from **A** to **B**.

   b) Lay the straight edge of a sheet of paper along the line of section.

   c) Mark onto the paper the start (**A**) and finish (**B**).

   d) Now carefully mark on the heights of the land (with a small line on the edge of the paper). To do this, mark the contour lines (notice that each 500 m contour marks the edge boundary of two types of dark shading) and nearby peaks with spot heights.

   e) Now draw axes onto a sheet of graph paper.

   f) Place your sheet of paper on the horizontal axis and carefully mark off the heights as crosses onto your graph.

   g) Join the crosses with a freehand pencil line.

   h) Use the map to add the names of the peaks.

   i) Complete your cross section by writing a title.

   j) Write a sentence comparing your cross section with the photograph of the mountain (Figure 17).

ICT ACTIVITY

Find out more about the eruptions of Mount Etna. Use the internet to focus on the following questions:

- What were the effects of the massive eruption in 1669?
- How in 1983 did people try to divert the flow of lava following an eruption?
- What eruptions have occurred since 2000 and have any of them caused damage to people and property?

Illustrate your answers with photographs if possible.

A Google search will reveal some good websites, but try:

Volcano World at http://volcano.und.edu

Wikipedia at http://en.wikipedia.org/wiki/Mount_Etna

F Issue: How should Abondance respond to the threat of global warming?

Abondance is a typical Alpine ski resort in the Haute-Savoie region of France (Figure 19). It is one of several traditional ski resorts in the area, depending on income from winter skiers for the survival of its hotels, restaurants and shops (Figure 20).

In 2007, following 15 years of making a loss, the ski lifts closed for the last time. The reason for the closure is due to a lack of winter snow caused, so scientists believe, by the impact of global warming. Abondance is not alone. Several other nearby resorts including Chatel are on the brink of survival. How can these resorts survive in a warmer world?

The French Alps is extremely popular as a tourist destination and this is likely to be the main source of income in the future. Farming is not very profitable and there are no major industries. The local council in Abondance is considering two options:

- To develop other forms of winter sports, such as ski touring, snow-shoeing (Figure 21) and snow-mobiling, that are less dependent on deep snow than traditional skiing.

- To develop its summer programme of activities to include hiking, water sports and mountain biking. This would enable it to become more of an all-year-round resort, rather than just a winter resort.

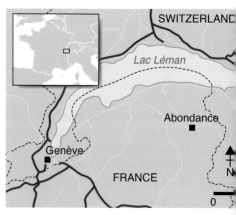

▲ Figure 19   Location of Abondance, France

▲ Figure 20   Abondance

▼ Figure 21   Snow-shoeing in the French Alps

RESEARCH

The aim of this activity, which will require some internet research, is to suggest a plan for the future of Abondance. You should work in pairs or small groups for this activity. What do you think the council should do to ensure the economic survival of Abondance in a warming world?

Use the internet to assess the options of expanding the winter sports programme or further developing plans for summer activities. Consider the best solution for the long-term survival of the town. When you have considered the options, decide on your recommendations.

Present your recommendations in the form of a Powerpoint presentation or booklet using Publisher. Remember that your recommendations must ensure a sustainable future for Abondance.

The following websites will get you started:

www.chaletinthemountains.com/sports.htm

www.simplysavoie.com

www.alpzone.com/snowshoe.html

www.valdabondance.com/IPsy.jpg

Coasts

A European coastlines

The coastline of Europe is incredibly varied. There are stretches of dramatic cliffs (Figure 1), wide sandy beaches (Figure 2) and bleak and desolate mudflats. Some of the coast is wild and remote with little sign of human activity. Elsewhere the coast has been developed with roads, railways, seaside resorts and busy industrial ports (Figure 3).

In this chapter, we will study some of the different types of coastline found in Europe and we will consider some of the issues facing coastal management today.

Why is there an isolated pillar of rock here?

Why are the cliffs so steep?

Why are the waves so powerful?

Why is the sand black?

▲ Figure 1   The coast in Iceland

▲ Figure 2 The coast in Denmark

Look at Figure 1. It shows part of the spectacular coastline of southern Iceland. When studying a stretch of coastline, it is interesting to pose some questions. These questions provide a focus for further investigation and help to improve our powers of observation and enquiry.

To answer the questions, we would ideally wish to visit the coastline to see it for ourselves. We could use books or the internet to help us identify and explain the formation of landforms. We could ask experts for their help or we could come up with our own ideas using our common sense.

▲ Figure 3 The coast in Spain

ICT ACTIVITY

Use the internet to find two photographs of European coastlines. One should show a natural unspoilt stretch of coastline, and the other should show a coastline that has been developed. A simple Google search 'Europe coast photo' will reveal many interesting sites and loads of possible photographs.

For each photograph, write a few sentences or add some labels to describe why it appeals to you. This could be done electronically, or by printing the photographs and sticking them into your book.

Alternatively, the photographs could be used to create a whole-class wall display, with each photograph being linked to a large map of Europe.

Activities

1 Study Figure 2.

a) Suggest some questions that you would be interested to investigate on this stretch of coastline. The questions on Figure 1 will help you.

b) Choose one of the questions you have posed and suggest how you might go about trying to answer it.

c) Imagine that you could visit either this stretch of coastline or the one shown in Figure 1. Which would you choose to visit and why?

2 Study Figure 3.

a) Suggest some of the different land uses on this stretch of the coast.

b) Why do you think people have developed the coast in this way?

c) What do you think the coastline was like before people developed it?

d) Copy the following opinion lines into your book. For each line, place an 'X' to indicate your opinion. For example, if you think the coast in Figure 3 is 'interesting', place your cross at this end of the line. The more 'interesting' you think it is the closer to that end of the line you should place your cross.

Interesting _____ Boring
Colourful _____ Dull
Peaceful _____ Noisy
Clean _____ Dirty
Fresh air _____ Polluted air

e) Would you like to visit this coastline? Explain your answer.

f) Are there any ways in which this stretch of coastline could be improved?

B The Green Bridge of Wales, Pembroke, Wales

Some of the most spectacular coastal scenery in the UK is found on the Pembrokeshire coast in southwest Wales (Figure 4). The magnificence of the landscape has led to the Pembrokeshire Coast becoming one of the UK's original ten National Parks.

Study Figure 5. It shows one of the Pembrokeshire coast's best-known landforms, the Green Bridge of Wales. In the past the Green Bridge of Wales formed part of a **headland** jutting out into the sea. Erosion of the headland at weak points in the rock formed **caves** (there is a present-day cave in Figure 5). The caves became larger and eventually the sea broke through the headland to form an **arch**. Further erosion and weathering weakened the roof of the arch, which eventually collapsed to form a **stack**. The low rocky outcrops you can see in the photograph are called **stumps**. They are the final remnants of stacks and are only exposed at low tide.

Notice in the distance the steep cliffs that mark the edge of the land. These cliffs are battered by powerful waves driven by winds blowing several thousand kilometres over the Atlantic Ocean (Figure 4). The waves **erode** the base of the cliffs (see Figure 6), undercutting them and maintaining their steep profile.

▲ Figure 5 Green Bridge of Wales

▲ Figure 6 Wave erosion

▲ Figure 4 Pembrokeshire coast

56

Activities

3 Study Figure 5.

   a) What coastal feature on Figure 5 do you think is the 'Green Bridge of Wales'?

   b) Do you think this is a good name for this feature? Explain your answer.

   c) This photograph was taken from a footpath that runs along the coast. Why do you think the footpath is very popular with visitors?

   d) Can you suggest why the Green Bridge of Wales is a popular image on postcards available in local shops?

   e) A National Park is an area of great natural beauty. Do you think it was a good idea to include this stretch of coast within the Pembrokeshire National Park?

4 Study Figure 7. It is a sketch of the headland that now forms the Green Bridge of Wales. Notice that the stack in Figure 5 has not yet been formed.

   a) Make a careful copy of Figure 7.

   b) Add the following labels to your drawing using Figure 5 to help you:
   - crack in the cliff
   - cave
   - arch
   - cliff

   c) Now draw one or more simple sketches of the headland, to show how it has been eroded to form the present-day headland in Figure 5. Use annotations (detailed labels) to describe what is happening.

ICT ACTIVITY

The Pembrokeshire National Park website (www.pcnpa.org.uk) has some excellent material, including some pages dedicated to the study of coasts at Key Stage 3.

You have been asked by the National Park Authority to produce a brochure for schools to encourage them to visit the coast for a field visit.

Access the section on coasts at www.pcnpa.org.uk/website/default.asp?SID=513&SkinID=2. Look at the photographs and, with the aid of the map, (www.pcnpa.org.uk/PCNP/live/sitefiles/related_items/pcnp_a4map.pdf), select three or four sites to visit that are all quite close together. School parties will not want to spend hours on a minibus or coach!

Now, plan the design of your brochure. It should include the photographs and some written information about each site. Try to use your own words. You should also include a map showing the location of your chosen sites. Think of a 'catchy' heading and make sure that your front page is interesting to look at.

When your class has produced the brochures, display them on the wall. You could ask someone to judge them and award a prize to the best one.

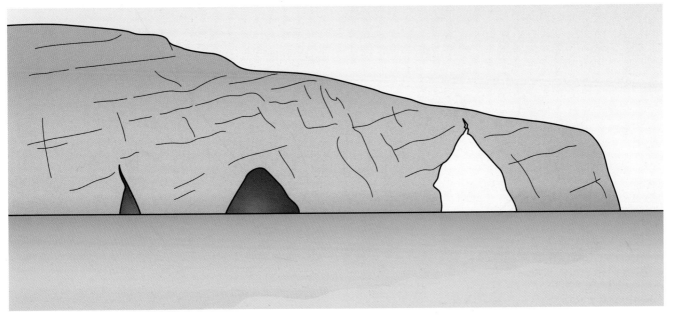

▲ Figure 7   Green Bridge of Wales (past)

C Fjords and killer whales: Norway's rugged coastline

Look at Figure 8. It is a satellite image of part of the northwest coast of Norway. Notice the jagged coastline and the snow-covered land. A map extract of this part of the coastline is shown in Figure 9. Locate the Lofoten Islands on both the satellite image and the map extract.

Activity

5 Study Figures 8 and 9.
   a) What is the name of the wide body of water at **A**?
   b) What are the names of islands **B** and **C**?
   c) Is the town of Narvik at location **D**, **E** or **F**?
   d) What is the name of the sea off the coast of Norway?
   e) Do you think this stretch of coastline is easy or difficult for ships to navigate? Explain your answer.

Notice on Figure 8 that there are several wide river valleys opening into the sea. These are called **fjords** (Figure 10). They are extremely deep and steep-sided flooded valleys formed by enormous glaciers during the Ice Age. Figure 11 describes how they were formed.

Much of the coastline is incredibly rugged and inaccessible. In the Lofoten Islands (see Figures 9 and 12, on page 60) the small coastal communities depend upon fishing. Their brightly coloured wooden houses, often built on stilts, line the seafront (Figure 12). Fish is often preserved by being salted and dried in the open air, rather like clothes hanging from a clothesline. It is a tough life in this wild, but beautiful, coastal environment.

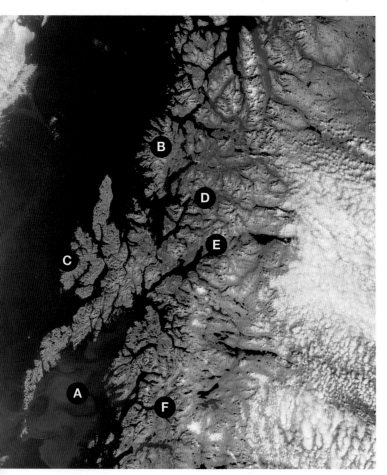

▲ Figure 8 Satellite image of NW Norway

▲ Figure 9 Atlas map extract of NW Norway

Activity

6 Study Figures 10 and 11.

   a) Describe the shape of the fjord shown in Figure 10.

   b) Why do you think many people choose to visit Norwegian fjords?

   c) Most people who visit the fjords do so by boat. Why do you think this is so?

   d) Use Figure 11 to help you explain in your own words how the fjord in Figure 10 was formed. Draw diagrams to support your explanation.

▲ Figure 10   Fjord in north west Norway

A During the Ice Age

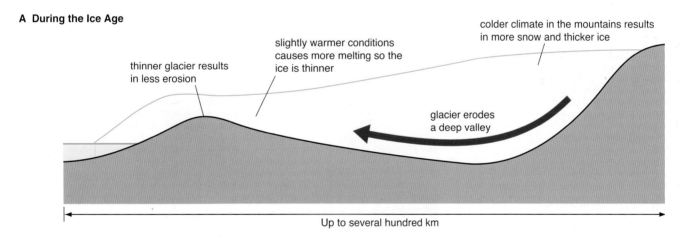

colder climate in the mountains results in more snow and thicker ice

slightly warmer conditions causes more melting so the ice is thinner

thinner glacier results in less erosion

glacier erodes a deep valley

Up to several hundred km

B After the Ice Age

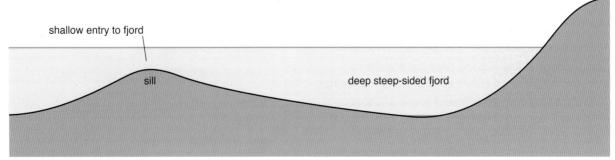

shallow entry to fjord

sill

deep steep-sided fjord

▲ Figure 11   Formation of fjord

Tourism has become popular in the Lofoten area. Some of the traditional fishermen's houses have been converted into rental properties, and there is much to see and do in the region. One of the great natural attractions of the area is the killer whales (Figure 13), which can be seen in some of the deeper fjords, such as Tysfjord.

▲ Figure 12   Coastal settlement on Lofoten Islands

Activity

7 Study Figure 12. The climate in this part of Europe is quite extreme. It is often cloudy and there is a great deal of rain. Winters are long and dark and cold.

a) Why do you think the houses are so brightly coloured? Do you think this is a good idea?

b) Why do you think some of the houses have been built on stilts?

c) Several houses in the photograph are rented out as holiday homes in the summer. Why do you think people choose to visit the area?

d) What do you think visitors would do here?

e) Apart from tourism, how do you think local people make a living?

▲ Figure 13   Killer whale calf in Tysfjord, Norway

ICT ACTIVITY

The Norwegian Tourist Board wants to promote northwest Norway for tourism. You have been asked to produce a single-sided page for a website, describing the attractions of the Lofoten area. You should include some text and some photographs. Include links to other websites that you think are informative. Design your webpage to look interesting and attractive. The following websites will help you to get started:

www.photomediaservice.com/norway

www.photographersdirect.com/stockimages/o/orca.asp

www.lofoten-info.no

D Coastal deposition: the Curonian Spit, Lithuania

The Curonian Spit is a 98-km-long, curved sandy ridge that stretches across a wide bay between Russia and Lithuania (Figures 14 and 15). On one side of the spit is the Baltic Sea, and on the other, the Curonian Lagoon. At its northern end, there is a narrow strait between the tip of the spit and the port of Klaipėda. The Curonian Spit is one of the largest features of coastal deposition in Europe and it has some of Europe's largest sand dunes, rising to over 60 m.

How do you think the spit has formed? According to local Baltic mythology a strong girl called Neringa created the spit while playing on the beach. An alternative theory suggests that the spit started to form about 5,000 years ago as huge quantities of sand were moved along the coast from the southwest and dumped at the end of the Sambian peninsula (Figure 14). Since then the sand has been carried northwards by the sea, cutting off the lagoon and forming the present-day spit. The movement of sand along the coast is called **longshore drift**. Look at Figure 16 on page 62 to see how it works. When waves approach the coast at an angle the **swash** carries pebbles up the beach. The **backwash** drags the pebbles back down the steepest part of the beach. In this way, pebbles move in a zigzag manner along the beach. This is longshore drift.

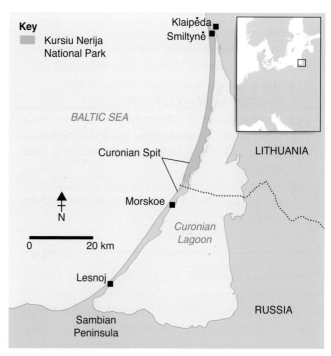

▲ Figure 14 The Curonian Spit

▲ Figure 15 Curonian Spit

Activity

8 Study Figures 14 and 15.

a) Is the town of Morskoe in Russia or Lithuania?

b) What is the name of the town at the northernmost tip of the spit?

c) How does the width of the spit change from south to north?

d) How many kilometres is it across the Curonian Lagoon at its widest point?

e) What is the name of the town shown by the letter **A** on Figure 15?

f) How do you explain the different shades of green on Figure 15?

g) Why do you think the colour of the water is different on either side of the spit?

h) What do you think the white areas are on the spit?

Activity

9 Study Figure 16. It shows how the process of longshore drift operates along this stretch of coast.

   a) Make a careful copy of Figure 17, which shows a stretch of coastline with a spit.

   b) Draw a large bold arrow to show the direction of longshore drift.

   c) Draw a series of zigzag arrows to show how sand particles are moved along the shore. Use Figure 16 to help you add labels to your arrows.

   d) What do you think will happen to the spit in the future? Show this on your diagram and add a label.

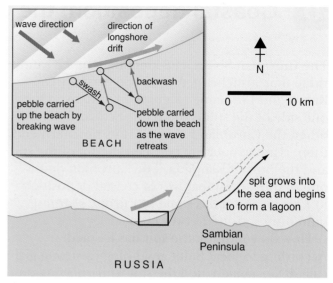

▲ **Figure 16 Longshore drift forming the Curonian Spit**

Plants soon colonised the Curonian Spit and dense forests grew, stabilising the sand and stopping it blowing away. However, in the Middle Ages, many trees were felled for building boats. Overgrazing by animals led to much of the vegetation being stripped. The sand was exposed and blew around, forming enormous sand dunes and burying villages. To halt the problem, a programme of tree planting began in 1825.

Today, the spit is highly valued as a thriving natural habitat for a great variety of plant and animal species (Figure 18). In 1991, the northern part of the spit became the Kursiu Nerija National Park. In 2000, the spit became a UN World Heritage Site, the same status as Stonehenge (which we studied in Book 1).

▲ **Figure 18 Curonian Spit**

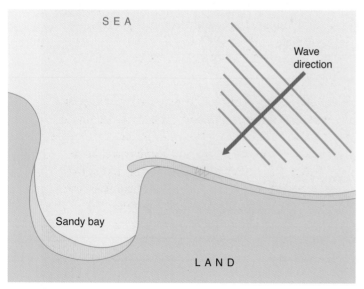

▲ **Figure 17 Spit in the past**

Activities

10 Study Figure 18.

a) Is the spit flat or hilly?

b) What is the evidence that the spit is made of sand?

c) Describe and suggest reasons for the location of the forest.

d) What evidence is there that the water in the photograph is the lagoon rather than the open sea?

e) The spit is popular with tourists. What are the attractions of the spit to tourists?

f) Would you like to visit the spit? What would you do there?

11 Study Figure 19. It lists some of the activities that are forbidden in Kursiu Nerija National Park, at the northern end of the spit (Figure 18). You have been asked to design a public information board at the entry to the Park to inform people about these rules. People visit the park from all over Europe so you need to use pictures and diagrams rather than writing. Design your information board on a sheet of plain paper. You can use ICT if you wish. Make your sheet as clear and colourful as you can.

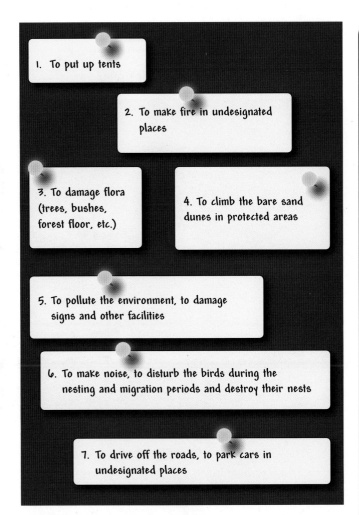

▲ **Figure 19   Forbidden activities in Kursiu Nerija National Park**

 ICT ACTIVITY

- The aim of this activity is to find a photograph of the Curonian Spit that can be labelled to show its main features. Use the internet to conduct a search of photographs and then select one that you think shows the main features of the spit (e.g. the sea or lagoon, the sandy beach, the high sandy ridge in the middle of the spit, the forests). Save your picture and then use the 'Draw' program to add some annotations (descriptive labels) describing the features. Print out your photograph and display on a classroom wall with others from your class.

The following websites will get you started:
http://en.wikipedia.org/wiki/Curonian_Spit
www.nerija.lt/en Kursiu Nerija National Park
http://photos.igougo.com/pictures-l8680-s2-Curonian_Spit_photos.html
www.worldheritagesite.org/sites/curonianspit.html

- There are some very interesting travel blogs on the internet where visitors have described what it is like to walk along the spit. Conduct a Google search 'curonian spit' to find some travel blogs. The World Heritage site (see above) has some interesting accounts. Make a list of some of the words or phrases that people have used to describe the spit. Does it make you wish to visit it? Explain your answer.

E Issue: Should an artificial surfing reef be constructed at Newquay, UK?

In 2008, Europe's first artificial surfing reef was constructed along Boscombe seafront, near Bournemouth in Dorset. Built at a cost of £1.5m, the reef is expected to attract up to 10,000 surfers a year. The artificial reef is constructed of sand-filled geotextile bags submerged on the sea bed. Waves approaching the shore are forced to break as the sea shallows over the reef, forming surfing waves up to 4 metres high.

A second European artificial reef has been planned for Newquay in Cornwall (Figure 20), but planning permission has yet to be granted. At a cost of about £6m, the artificial reef could generate up to £1.6m a year from the increase in the number of surfers to the town (Figure 21). However, there are strong arguments both for and against the proposal, as you will discover.

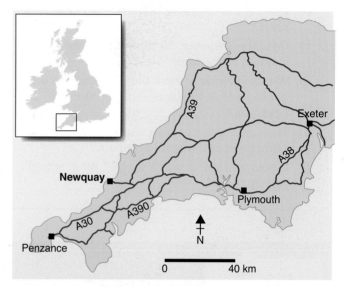

▲ Figure 20   Location map of Newquay

▲ Figure 21   Surfers at Newquay

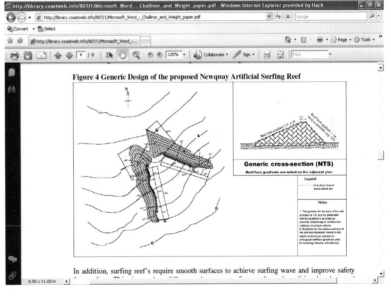

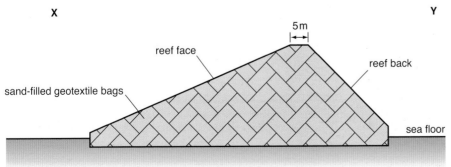

▲ **Figure 22   The artificial reef scheme at Newquay**

Activities

12 This activity will take the form of a Public Enquiry involving the whole class.

a) Divide up the class to represent the following interest groups:

In favour of the scheme:

- South West Tourism
- Tunnel Vision Surf Shop
- Cornwall Wildlife Trust
- local restaurant owners (for example, celebrity chef Rick Stein, who would be tempted to open a restaurant in Newquay).

Against the scheme:

- Newquay Rowing Club
- fishermen
- local environmental group, concerned about the impact of the reef on natural coastal processes and wildlife
- local people concerned about a huge influx of people into the town (parking could be a real issue).

You also need a panel of judges (say 3–4) who will preside over the Enquiry.

b) Each interest group needs to conduct some background research to put forward their case to the judges. Look at Figures 22 (page 65) and 23 and use the internet to help. The judges need to have an overview so should consider the merits of all arguments both in favour of and against the scheme. The judges should come up with some questions to ask the interest groups.

c) Sort the statements in Figure 23 to identify which are advantages and which are disadvantages.

d) To run the Public Enquiry one of the judges needs to set the scene. Then the views of the interest groups should be heard. Start with those in favour and then hear those against. Follow the presentations with questions from both the interest groups and the judges.

e) Finally the judges need to make a decision.

13 Consider the outcome and discussions in your class's Public Enquiry.

a) What was the judges' decision?

b) Do you agree with the decision? Explain your answer.

c) What do you think was the most powerful argument in favour of the scheme? Why?

d) What do you think was the most powerful argument against the scheme? Why?

Local restaurant owners would benefit from more people and from the lobsters produced on the reef.

The reef will alter the natural coastal system, creating new powerful currents and interfering with sediment movement. Erosion might be increased elsewhere along the coast.

Sand builds up on the sheltered side of the reef. This widens the beach and helps to protect the coastline from harmful erosion.

It is expensive. The construction of the reef would cost about £6m.

Local people are concerned about an influx of more surfers. They are concerned about the lack of parking in the area.

Fishermen are concerned that the reef could make entry into the harbour for trawlers more difficult due to the creation of powerful currents.

Newquay Rowing Club is concerned that the project would ruin a 100-year-old triangular racing route in the bay.

It is estimated that the reef would generate £1.6m a year in boosting local surf-related industries and surf shops.

Hotels would benefit, as the artificial reef would produce more days of surf, attracting visitors throughout the year.

Biodiversity increases as marine organisms colonise the reef. At Newquay there are plans to plant a field of kelp (large brown seaweed that provides shelter for plants and animals) and to construct a lobster hatchery.

▲ Figure 23 Advantages and disadvantages of building an artificial reef at Newquay

People

CHAPTER 5

In this chapter you will study:

- celebrating diversity in Europe)
- population distribution in Europe
- migration into southern Europe from Africa
- Europe's ageing population
- should Europeans have larger families?

A Multicultural Europe

▲ Figure 1   People in Europe

Europe is a continent rich in different traditions and cultures. Over many centuries, the history of European trade with other parts of the world has brought together a tremendous range of peoples and cultures (see Figure 1). Despite the many wars of the past, Europeans now try to live peacefully with one another, solving their disputes round a conference table rather than on the battlefield.

Not only do the people of Europe vary in their languages, physical features and in the clothes they wear, but they also have different cultures, beliefs and interests. For example, a large number of people are Roman Catholics, Orthodox Christians, Protestants and Muslims – in fact, many of the world's religions are represented across Europe.

We can see this variety in our restaurants, shops and sports (Figure 2), and in different styles of art, music and architecture across the continent (Figure 3). We need to continue to celebrate this diversity and work together for the future benefit of Europe and the world.

▲ Figure 2   Food, fashion and sport across Europe

'Self-portrait',
Van Gogh

Gaudí architecture, Barcelona

Sigur Ros, Iceland

▲ **Figure 3 Artistic diversity across Europe**

Activities

1 For this activity you can either work individually or in pairs. Consider the impact of European cultures and traditions on your life.

   a) Can you think of any foods that have been introduced into the UK from other European countries? Name the foods and the countries.

   b) In 2008, the famous Tour de France cycle race included a stage in Italy. In 2007, there was a stage of the race in the UK. Make a list of sports that involve a number of different European countries.

   c) Now consider television and film. Can you think of any TV programmes or films that have involved places or people from other European countries?

   d) Finally, the arts! Can you think of any painters, authors or architects that come from European countries? Have you read any books written by Europeans outside the UK?

2 The aim of this activity is to create a wall display in your classroom entitled 'Celebrating diversity in Europe'. To do this you need to find pictures illustrating as many features and characteristics of people's lifestyles, cultures and traditions across the continent. Consider using food labels, pictures and photographs from the internet, magazines and tourist brochures. Try to cover as many aspects as you can, including the arts, fashion, food, music, sport and architecture. Remember that this display is about people rather than natural landscapes.

ICT ACTIVITY

Study Figure 4. It shows some of the lesser-known European sports. Unfortunately the labels have all been mixed up!

- Can you sort them out? Make a list of the sports and write the correct number alongside each one.
- Choose one of the sports and use the internet to find out a bit more about where it is played in Europe and what it involves.
- Do you think your chosen sport would be popular at your school? Explain your answer.

Basque Pelota (Spain)

Hurling (Ireland)

Handball (Germany)

Bocce (Italy)

Petanque (France)

Kiiking (Estonia)

Bandy (Russia)

Curling (Scotland)

Korfbal (Netherlands)

▲ **Figure 4   Examples of lesser-known sports in Europe**

B Europe's population distribution

Look at Figure 5. It is a satellite photograph showing Europe at night! As you can see, much of Europe is giving off light (shown blue on the satellite image).

Light is a good measure of the spread or **distribution** of population. Can you see the concentration of light marking major cities such as Paris in France and Madrid in Spain? Notice that the countries of Western Europe (such as Germany, Italy and the UK) are giving off a great deal of light. These countries have high populations with many people living in well-lit towns and cities. The darker parts of Europe are towards the east, including the Baltic states of Estonia, Latvia and Lithuania (which are countries with smaller populations and fewer towns and cities).

Activity

3 Study Figure 5.
   a) Use the atlas map in the inside back cover to help you identify the cities **A** to **E**.
   b) Locate the UK. Describe the pattern of the brightest lights in the UK.
   c) Where are the darkest areas in the UK?
   d) Why do you think these darker parts of the UK have fewer people living there?
   e) Locate Italy. Describe and try to explain (see Figure 1, page 40) the pattern of light in Italy.
   f) Not all of the light on Figure 5 comes from land sources such as streetlights. Notice that there is a line of light sources running through the North Sea (**X** on Figure 5). Can you suggest the origin of these lights? (Hint: think what is extracted from the North Sea)
   g) Satellite images such as Figure 5 are less useful in showing population distribution in poorer parts of the world. Can you suggest why?

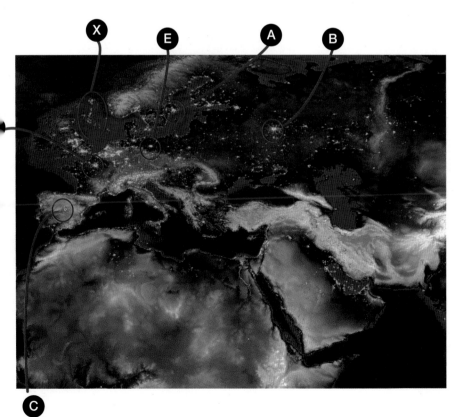

▲ **Figure 5 Satellite photo of Europe at night**

RESEARCH

In this activity you are going to try to explain the pattern of light (population distribution) in Sweden. Locate Sweden on Figure 5, page 71. Notice that the brightest lights are in the south of the country. This is where most people live. Your task is to try to find out why.

In Book 1, you learned that population distribution was affected by positive and negative factors (Figure 6). You need to try to investigate the factors listed in Figure 6 to see if you can explain why most of Sweden's population live in the south of the country. Here are some suggestions:

- Make use of atlas maps from your Geography room or the library. They contain maps showing climate, land use, communications, etc

- About Geography is a good place to start on the web at

http://geography.about.com/library/maps/blsweden.htm

- There is useful information in the CIA World Factbook at
http://geography.about.com/library/cia/blcsweden.htm

- Wikipedia is at http://en.wikipedia.org/wiki/Sweden

- The US Department of State has a useful site at
www.state.gov/r/pa/ei/bgn/2880.htm

- Information about Sweden's climate can be found at the Met Office's website at
www.metoffice.gov.uk/weather/europe/sweden_past.html

This is your research, so you make the decisions and see what you can find out! The short extracts in Figure 7 will help you get started. Good luck!

Positive factors encouraging people	Negative factors discouraging people
Flat land is easy to build houses, industry and roads on	Steep land and mountains are hard to build on and make communications difficult
Fertile land encourages farming	Poor-quality soils have few uses apart from forestry and livestock grazing
Natural harbours for ports to develop; trade with other countries will encourage industry and provide employment for people	Lack of natural harbours or ports results in the area being cut off from trade with other countries
Raw materials for energy or industry encourages the development of industry	Lack of raw materials for industry and energy
Moderate climate with no extremes of temperature and rainfall	Extreme climate, e.g. cold winters with snow
Near to other countries to encourage trade and economic links	Remote from other countries

▲ **Figure 6 Factors affecting population distribution**

Sweden has the largest population of the Nordic countries. It is separated in the west from Norway by a range of mountains. It shares the Gulf of Bothnia at the north end of the Baltic Sea with Finland.

The southern part of the country is chiefly agricultural, with forests covering an increasing percentage of the land the further north one goes. Population density is higher in southern Sweden.

Sweden, which occupies the eastern part of the Scandinavian Peninsula, is the fourth-largest country in Europe. The country slopes eastward and southward from the Kjølen Mountains along the Norwegian border. In the north are mountains and many lakes. To the south and east are central lowlands and south of them are fertile areas of forest, valley and plain.

▲ **Figure 7 Information about Sweden**

There are almost 500 million people living in the 27 countries of the European Union. The people of Europe are, however, very unevenly spread across the 27 countries.

The country with the highest population is Germany, with 82.4 million people. Look back to Figure 5 to see how bright Germany is at night! France, the UK and Italy have populations close to 60 million. The countries with the smallest population are Cyprus, Luxembourg and Malta, each with less than 1 million people. Often, the larger the country, the higher the population.

Whilst Germany is the most populous country in the EU, it is not the most crowded. Despite having the smallest population, Malta is by far the most densely populated country with 1261 people per square km! This compares with the UK figure of 246 people per square km (Figure 8).

Activities

4 Study Figure 8. In this activity you are going to produce a special type of map called a **choropleth** map. Geographers commonly use this type of map. We are going to map the population density for the 27 EU countries. To complete the map, you need to work through the following steps using a blank outline map of Europe.

a) A choropleth map uses a sequence of progressively darker colours to show increasing values. The values are grouped together into four to eight categories. It is important to make sure that the categories do not overlap. Write the following categories using the correct colours on your map to form a key.

Population Density (people/km²)	Colour
251+	Black
201–250	Brown
151–200	Dark red
101–150	Light red
51–100	Orange
50 and below	Yellow

Notice that the colours merge into each other. This makes the map easier to interpret.

b) Now shade each country according to the key using the values in Figure 8. (For example, Germany has a population density of 231 people km². This means that it should be shaded brown.)

c) When your map is complete, label a few of the countries (particularly those with the highest and lowest population densities) – you may not have room to label all 27 countries!

d) Now give your map the following title: 'A choropleth map showing the population density of EU countries'.

e) Compare your map with Figure 5. Are the countries that are brightly lit up at night, the ones with the highest population densities?

Member State	Population density (People / km²)
European Union	112
Austria	99
Belgium	344
Bulgaria	70
Cyprus	84
Czech Republic	131
Denmark	126
Estonia	29
Finland	16
France	99
Germany	231
Greece	84
Hungary	108
Ireland	60
Italy	195
Latvia	35
Lithuania	52
Luxembourg	181
Malta	1,261
Netherlands	394
Poland	122
Portugal	114
Romania	91
Spain	87
Slovakia	111
Slovenia	99
Sweden	20
United Kingdom	246

▲ Figure 8   Population densities of countries in the European union

C African migration into Europe

Look at Figure 10. In the background, notice the sandy beach with sun lounges and parasols. It is a popular tourist beach on the Canary Isles, a group of islands in the Atlantic Ocean just off the coast of northwest Africa (Figure 11).

Now look in the foreground of the photo and notice the large open-topped boat crammed full of people. These are migrants from Africa who have made the perilous journey to escape poverty or persecution. They are in search of jobs and a better quality of life in a European country. For them the Canaries, which belong to Spain, are seen as a nearby entry point to Europe.

In recent years, tens of thousands of people have made the dangerous journey to the Canaries. They often arrive on the island's beaches dehydrated and exhausted (Figure 12). Many thousands are thought to die en route, as their boats are frequently in poor condition.

In the past, some of the migrants have managed to gain entry into Spain where they have found work and are able to enjoy a higher standard of living than they did in Africa. This has been the attraction to people trying to escape the poverty of Africa.

More recently, Spain and other European countries have tightened up their borders, making it much harder for illegal immigrants to enter. European Union joint border guards (called Frontex) patrol the borders to try to intercept the African boats and their human cargo. Those who are caught are usually sent home unless they can prove that their lives are in danger.

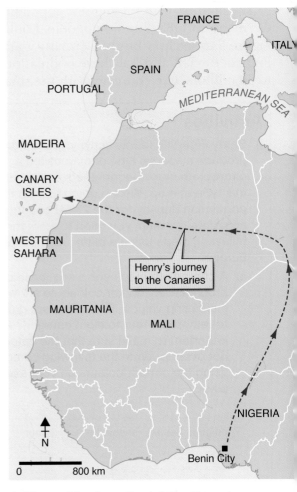

▲ **Figure 11   Location of the Canaries**

◀ **Figure 10   African migrants**

Activities

5 Study Figure 10.

   a) What are the people doing on the beach in the distance?

   b) Estimate the number of African people in the open boat.

   c) Suggest some of the possible dangers for the African people, as they travel across the Atlantic to the Canary Islands.

   d) What do you think is happening to the African boat and the people in it?

   e) Why do you think the African people want to enter Europe? Think of at least three reasons.

   f) Do you think the people in the boat should be allowed to stay in Europe or should they be sent home? Give reasons for your answer.

6 Read Figure 13, which is a personal account of a migrant's journey from Nigeria to the Canaries.

   a) Where did Henry come from?

   b) Why did he want to leave his home and move to Spain?

   c) Make a copy of the map in Figure 11 showing Henry's route. Use an atlas to name the countries through which Henry travelled.

   d) Use the scale on Figure 11 to estimate how far Henry travelled through Africa before getting on the boat.

   e) The man who took money from Henry, in exchange for a place on the boat, is a smuggler. He is involved in smuggling people illegally from one country to another. These people often make a lot of money out of other people's misery. How much did Henry pay the man?

   f) Henry was one of 75 Africans put into three boats. How much money did the smuggler collect altogether? Use the internet to convert euros into pounds.

   g) What do you think probably happened to the other two boats that Henry didn't see again?

   h) Do you think Henry will try again? Explain your answer.

Henry's Story

"My name is Henry. I am 30 years old. I have tried to get to the Canary Isles once before, but my boat got intercepted by the Spanish police and I was sent back. It was the most frightening time of my life, although I am going to try again.

Life in Nigeria is hard. There is very little food and there are no jobs. Many people die in poverty. There is so much corruption too. I want to make a better life for myself in Spain and earn money to send back to my family.

My journey started in Benin City. I travelled overland through Nigeria, Niger Republic, Libya, Algeria and finally into Morocco. I worked on the way as a barber to earn some money.

In Morocco, a friend told me to go to the western side. Here I paid a man 300 euros to put me on a boat bound for Europe. There were 25 people in the boat. As we drifted in the ocean many people were sick. A big wave came over the boat and I thought I was going to die. The engine of the boat often stopped working and we drifted helplessly, with no idea where we were going.

The police picked us up just before our boat broke into two and sank. Two other boatloads of Africans left Morocco at the same time. I never saw them again.

Life is so terrible in Nigeria that I will try again – soon."

▲ **Figure 13 Account of African migrant**

◄ **Figure 12 Migrants on Canary beach**

D Europe's ageing population

Did you know that the fastest growing age group in the UK is the over 85s? Joan (Figure 14) is one of over 1 million people who are over the age of 85. She is, in fact, over 90! Over 10 million people are aged over 65. Do you know anyone who is over 85? How old are they?

There are several reasons to explain this **ageing** trend (Figure 15). Most of them are to do with the general improvement in living standards over the last few decades. Most elderly people now have access to good health care and are able to live healthy, active and fulfilling lives long after they have retired.

An ageing population brings with it a range of challenges and opportunities (Figure 16). In the future an increasing number of elderly people will need to be looked after. Some may need specially designed homes whilst others may need expensive health care. Some may need financial help to buy fuel. Others may need transport to go shopping or to the doctor's.

Despite the problems of an ageing population, older people have a huge amount to offer. They are very wise and knowledgeable and can use their experience to guide younger people. They are often very reliable and efficient in the workplace and many companies now prefer to employ 'retired' people than youngsters. As grandparents, older people are highly valued in families, often providing support when parents are at work.

Several other European countries are also experiencing an increase in the number of elderly people such as Italy, Germany and Norway. With more elderly people and fewer younger people there is concern about who will be earning the money to support the elderly in the future. This explains why some European countries have begun to encourage couples to have more children.

Joan is 91. Here she is in her garden in Scotland. It is her pride and joy. You wouldn't know it, but she is almost completely blind, yet she manages to tend her large garden and look after herself and her partner Peter in their house near New Galloway. She is very independent and enjoys cooking and making woollen sweaters. Her poor eyesight prevents her reading, but she can enjoy the enlarged photos given to her by her family. She has a special telephone with large numbers so that she can see to make calls. She is a much-loved member of a very big family.

▲ Figure 14 Joan

Why are more people living longer?

Living conditions have improved. Most people live in comfortable homes with central heating.

The health service, with new drugs and medicines, keeps people alive for longer.

Diets have improved. A balanced diet results in a healthy body.

More people were born just after World War I (1914 to 1918). Those born just after World War II (1939–1945) are nearing retirement now. This is the so-called 'baby boomer' generation.

Fewer people live in poverty. Pensions and other benefits help older people.

▲ Figure 15 Why people are living longer

▲ **Figure 16 Ageing population**

Set in the thriving location of Nunthorpe near Middlesbrough, Roseberry Mews will consist of 41 one-bedroom and 15 two-bedroom private apartments. Situated close to the picturesque Cleveland Hills, Nunthorpe is ideally placed for visiting the North Yorkshire Moors and the coast.

Nunthorpe benefits from a comprehensive parade of shops, a railway station and other public transport links. All are within walking distance of the development. Roseberry Mews is the perfect location to lead a comfortable, active retirement.

In the development, the emphasis is on security, safety, style and comfort. The development will have an estate manager along with CCTV, 24-hour Careline, communal lounge, lift and guest suite. Professionally landscaped gardens will surround the buildings, creating a beautiful environment to live in.

▲ **Figure 17 Roseberry Mews retirement apartments**

Activities

7 Study Figure 14. Think about an elderly person who you know. It could be a member of your family, for example a grandparent, or a friend of the family. Maybe you have an elderly teacher!

a) Write a few sentences to describe the life of your chosen person. When was he/she born? Where does he/she live? What job or career did he/she have? What does he/she do with his/her time now? Try to find a photo of your chosen person.

b) Suggest some of the special needs of your chosen person. It could be regular health visits or an exercise regime. It could be help climbing the stairs.

c) Do you think your chosen person has a good quality of life? How could his/her quality of life be improved?

d) How have you benefited by knowing your chosen person?

8 Study Figure 17. It describes a proposed retirement development near Middlesbrough in northern England.

a) Why do most of the apartments have only one bedroom?

b) What services are available to the residents within walking distance?

c) Elderly people are often concerned about security and safety. How are these two issues being addressed at Roseberry Mews?

d) What is being done to make it an attractive place to live?

e) Do you think anything else should be added to the development to provide residents with a high quality of life?

RESEARCH

In this activity you are going to conduct some research into a local retirement complex. You will then have the opportunity to design your own retirement complex!

1 Conduct a Google search to find out about a local retirement complex. Search for 'retirement home' or 'retirement development' and add your nearest town or district. Make some rough notes on the kinds of property available, the services provided to support the residents and what has been done to make the environment attractive. Add any other information that you think could help you in your own plans for a retirement development.

2 Plan a retirement complex in your local area.
- Decide where it will be located and why. Include a sketch map to show its location.
- What types of accommodation will you provide? Include some sketches or photos.
- Describe the layout of your development. Draw a simple plan to show where the buildings and facilities are located. Consider gardens, paths and car parks.
- Write a short description of your development (as in Figure 17) outlining its attractions for elderly people.

E Issue: How can Germany encourage a 'baby boom'?

Do your parents ever tell you how expensive you are to bring up? If so, it won't surprise you to learn that a recent trend in Europe has been for women to have fewer and fewer children.

Many women are deciding not to have children until they are older, so that they can concentrate on their careers. This means that they tend to have smaller families (Figure 18). Couples may choose to enjoy a high quality of life free from the burden and expense of children.

In Germany, 30 per cent of women have decided not to have children. For the future, this could cause problems for the German government. With fewer people entering the workforce, there will be less wealth creation. How will the ever-increasing number of older people be supported in the future?

Some European governments have introduced policies in an attempt to encourage families to have more children. These policies include paid maternity leave, improving day care for children and even paying families for each child born! In Germany mothers get 14 weeks' paid leave and there are tax incentives available for families with children. Despite these policies, Germany continues to have one of the lowest birth rates in Europe.

▲ **Figure 18 A small family in Germany**

Activities

For these activities you can work either individually or in pairs (ideally boy/girl).

9 Read the account below, which describes the problem in Germany.

Germany has long had one of the lowest birth rates in the European Union and one of the highest proportions of childless women. One of the biggest problems is a lack of child-care places. According to government figures, only one in five children under three get a place in day care. Not only do they close at lunchtime, but also the fees are incredibly high. Another problem for working parents is that, traditionally, the school day ends at 1pm.

a) Why do you think poor child-care provision could put German people off having children?

b) Why is it a problem for parents if schools close at 1pm?

10 In order to improve child care in Germany, imagine that schools have been asked to organise a programme of extra-curricular activities in the afternoons. This will allow parents to have a longer working day. Teachers and other professionals will supervise the activities.

a) Work in pairs or small groups to suggest a programme of activities for Monday to Friday from 1 to 4pm for pupils in the equivalent of either Year 1 or Year 6. You should have a good range of options to include academic (e.g. English speaking), sporting (e.g. football), leisure (e.g. cooking), arts (e.g. painting), performing arts (e.g. dance). Present your programme using a table in Word or Excel.

b) Design a single-sided advert to appear in the local newspaper advertising your 'Afternoon Activities Programme'. You must aim your advert at young couples who may be considering starting a family and who are concerned about child care as their children grow up.

European Cities

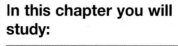

In this chapter you will study:

- the location of Europe's capital cities
- the development and characteristics of Tallinn, the capital of Estonia
- the issue of transport in European cities
- Green Cities – building for a sustainable future
- how to prevent flooding in Venice, Italy.

A European cities: an introduction

Did you know that three out of every four people in Europe live in a town or a city? The European city with the highest population is London, with a population of 7.5 million people. The next-largest city is Berlin, with a population of 3.4 million people.

Look at Figure 1, which lists the top ten most populous cities in the European Union. Use the atlas map (see inside back cover) to locate these cities in Europe.

Activities

1 Study the political map on the inside back cover.

   a) What symbol is used on the atlas map to show a capital city?

   b) Alphabet Run! Can you find four capital cities in Europe beginning with the letters **A** to **D**?

   c) One of the ten cities listed in Figure 1 is not a capital city. Which one is it?

   d) Rather strangely, one European country has two capital cities! Can you find the country and name the two capital cities?

2 Study Figure 2. Unfortunately the capital cities and the countries have been muddled up. Use the atlas map (inside back cover) to help you sort them to correctly match the capital cities to the countries.

3 For this activity you will need a blank outline map of Europe. You are going to produce a map to show the location and population of the top ten cities in the European Union.

   a) Use the atlas map (inside back cover) to locate each of the cities listed in Figure 1. Mark the locations with a pencil dot.

   b) For each city draw a bar to represent the population. A scale of 1 cm = 1 million should work well but you may wish to alter this according to the size of your map. Keep the width of your bars the same, for example 1 cm. Draw each bar vertically on the map. Locate the base of the bar as close to the location of the city as possible. Try to avoid overlapping.

   c) Now shade the bars using a single colour.

   d) Write the name of each city alongside the bar.

   e) Complete your map by adding a scale for the bars and writing the title on your map.

Rank	City/Country	Population (million)
1	London, UK	7.5
2	Berlin, Germany	3.4
3	Madrid, Spain	3.2
4	Rome, Italy	2.8
5	Paris, France	2.2
6	Bucharest, Romania	1.9
7	Hamburg, Germany	1.8
8	Warsaw, Poland	1.7
9	Budapest, Hungary	1.7
10	Vienna, Austria	1.7

▲ Figure 1 Top ten most populated European cities

Country	Capital city
Finland	Vienna
Lithuania	Bratislava
Czech Republic	Paris
Austria	Prague
Portugal	Helsinki
France	Vilnius
Denmark	Nicosia
Belgium	Lisbon
Slovakia	Brussels
Cyprus	Copenhagen

▲ Figure 2 European countries and capital cities (muddled up)

B Tallinn: a European capital city

▲ **Figure 3 Toompea Hill, Tallinn**

Tallinn is the largest city and capital of Estonia. The city lies on the north coast of Estonia and is separated from Helsinki, the capital of Finland, by the Gulf of Finland.

The history of Tallinn

Tallinn is a very ancient city. It began as a trading settlement around AD800, making use of its coastal location. The early settlement was sited on top of Toompea Hill (Figure 3). This site was relatively easy to defend because it was possible to keep a lookout over the surrounding countryside. A castle was built on the hill and parts of it still remain today. In the Middle Ages, the settlement grew and expanded outwards from Toompea Hill, and Tallinn soon became a centre for trade and commerce.

Tallinn today

However, for most of its history, Tallinn has been the focus for wars and battles, as neighbouring countries have fought for control of the city and the country. The local people have suffered terribly from famines and plagues as the wars have raged. During the Second World War (1939 to 1945), the city was severely damaged by Russian bombers. After the war, Russia seized control of the country.

In 1991, Estonia achieved independence from Russia, and since then the city has changed dramatically. The old Soviet factories and poor-quality housing estates (Figure 4, page 82) have been replaced with modern developments. The Old Town has been restored (Figure 5, page 82) and the economy is now booming, led by high-tech companies and other modern service-sector industries. The population of the city has grown to about 400,000, which is about the same size as Bristol.

Tallinn has become a popular city for tourism, particularly since it has become linked to Western Europe by the budget airlines (such as easyJet). In the summer, the city is heaving with coach parties and independent travellers. About 3 million tourists visit the city each year. It is a lively and colourful city with shops, parks, museums and theatres. As a measure of its cultural importance, Tallinn has been nominated to be a European Capital of Culture in 2011.

Tallinn: capital city of Estonia

Capital cities have a range of important functions that make them special. Just like London in the UK, Tallinn is the centre for government, finance, business, transport, education and culture. This is what makes a capital city different from other large cities.

Estonia's parliament is based in Toompea Castle in the centre of Tallinn and many government buildings are located in the city. Tallinn is home to a large number of foreign embassies, including the British Embassy. Tallinn is a centre for sport and the arts. It has a number of large sports stadiums, museums, theatres and concert halls. The Estonian National Library is also situated in Tallinn. In 2002, Tallinn hosted the Eurovision Song Contest!

Tallinn has a number of important universities and educational buildings, including the Estonian Business School, and the Academies of Music and Art. With its international airport and mainline rail station, Tallinn has become the main financial centre in Estonia, with headquarters of banks and businesses located in the city.

▲ **Figure 4   Tallinn during the Soviet era**

▲ **Figure 5   Old Town (restored), Tallinn**

Activity

4 Study Figure 5. The photograph was taken in the Old Town of Tallinn.

a) What time of year do you think this photograph was taken? Explain your answer.

b) What is the evidence that this street is lined with shops aimed at tourists rather than local people?

c) Do you know the name of the type of road surface that the people are walking on?

d) What is the evidence that this photograph was taken in the old historic part of Tallinn?

e) Do you think this street was designed to cope with cars and lorries? Explain your answer.

f) Most of Old Tallinn is pedestrianised. What do you think this means?

g) Notice that there are a few cars parked on the street. Do you think all cars should be banned from Old Tallinn? Explain your answer.

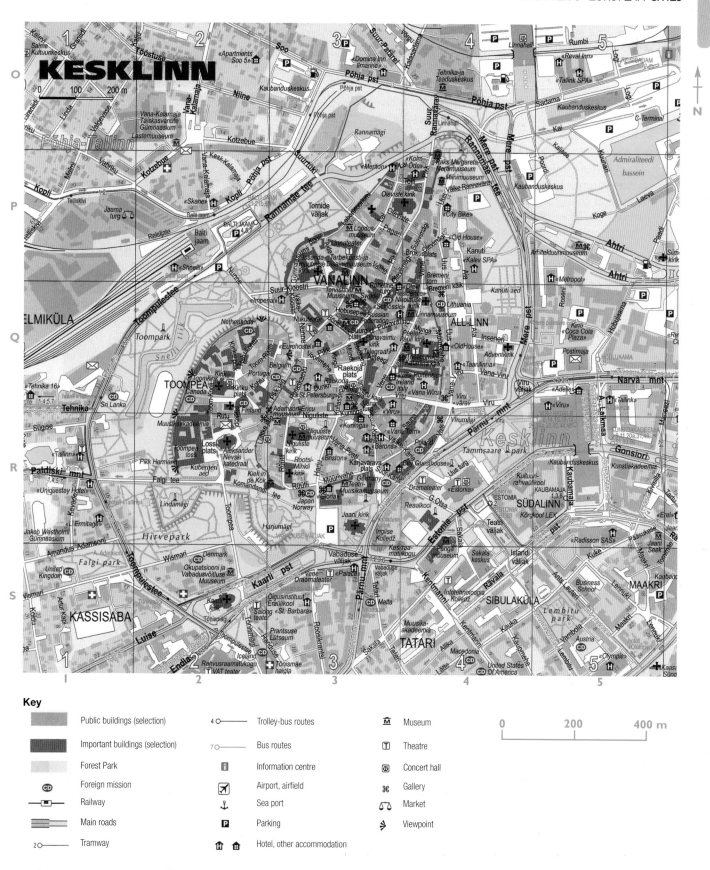

▲ Figure 6 Extract from city map of Tallinn

Key

Public buildings (selection)	
Important buildings (selection)	
Forest Park	
CD Foreign mission	
Railway	
Main roads	
2○— Tramway	

4○— Trolley-bus routes	
7○— Bus routes	
ℹ Information centre	
✈ Airport, airfield	
⚓ Sea port	
P Parking	
🏨 🏨 Hotel, other accommodation	

Ⓜ Museum	
Ⓣ Theatre	
Concert hall	
⌘ Gallery	
⚖ Market	
⤳ Viewpoint	

0   200   400 m

Activities

5 Study Figure 6 on page 83.

a) Many foreign tourists arrive in Tallinn by aeroplane or ferry. How else do you think tourists might travel to Tallinn? Give evidence from the map.

b) Locate the Old Town in Tallinn. It is the area inside the circular yellow ring road. Remember that Toompea (grid square Q2) was the original site of the settlement. Notice that there are several viewpoint symbols around the edge of Toompea. What does this tell you about the relative height of this district compared with the rest of the Old Town?

c) Many of the buildings in the Old Town are coloured purple. Use the key to discover what this means.

d) In what grid square is the Tourist Information centre?

e) How many museums are there in the Old Town?

f) There are several red symbols with the letters CD. What does this symbol mean? Why do you think there are a number of these in the centre of Tallinn?

g) Notice that there is a ring of parkland surrounding the Old Town. In which park is the lake Snelli tiik?

h) What are the advantages of having all this green space so close to the city centre?

i) Describe and suggest reasons for the location of the car parks on the map.

j) Notice that there are several hotels in the Old Town. The cost of staying in these hotels tends to be high compared with the rates charged for staying in hotels elsewhere in the city. Can you explain this?

6 Look again at the photographs and the map extract.

a) Why do you think Tallinn is a popular destination for tourists?

b) Would you like to visit the city? Explain your answer.

7 One of the reasons for Tallinn's recent expansion is because it is well connected with other European cities (Figure 7).

a) With reference to the atlas map (inside back cover) locate Tallinn on a blank outline map of Europe.

b) Now locate the cities listed in Figure 7, which have air connections with Tallinn.

c) Use a single colour, say red, to draw arrows to connect these cities to Tallinn.

d) How well connected is Tallinn to the rest of Europe?

e) Are certain parts of Europe better served than others?

f) Imagine that you are a wealthy owner of an airline. You can choose to connect one additional European city to Tallinn. Which would it be and why? Use a different colour to show this on your map.

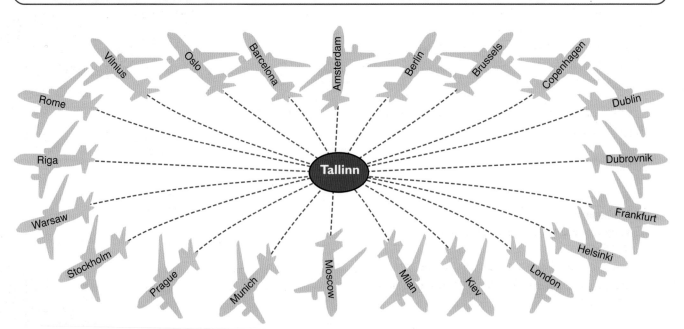

▲ Figure 7 Major European cities connected to Tallinn by air

ICT ACTIVITY

You and your family have decided to visit Tallinn for a short break during one of your school holidays. In this activity, you have to decide when you are going to visit Tallinn (during which school holiday), how you are going to get there, where you are going to stay and what you are going to do. Use the internet to help you conduct your research. Present your findings in the form of a short report (using photographs and maps if appropriate).

- When to visit? You have to stick to school holidays. Find out about the climate of Estonia – it is very cold in winter but can be wet and busy in the summer. Wikipedia has climate information at http://en.wikipedia.org/wiki/Tallinn

- How to get there? You need to look at your local airports and investigate flights with the budget airlines, such as easyJet, as well as the Estonian carrier Estonian Air.

- Where to stay? A Google search on 'Estonian hotels' will reveal lots of possibilities. Add 'trip advisor' to your search to find useful feedback from people who have actually stayed in the hotels.

- What to do? The section in this book will give you some ideas but you will find many others on the internet, such as www.tourism.tallinn.ee www.balticsww.com/tourist/estonia

This activity can be done either individually or in pairs. Conduct your own study of a European capital city. You can choose any city you like, apart from London and Tallinn. Your aim is to find out what makes your chosen city special. Find out what makes it unique and why people might be interested to visit it. For example, your title could be something like 'What makes Rome special?'. Present your information in the form of a short Powerpoint presentation, lasting a maximum of 10 slides.

C Transport in European cities

Did you know that on average a European citizen makes 1,000 journeys a year and the vast majority are less than 5 km in length? About 75 per cent of all journeys in towns and cities are made by car, often travelling at speeds less than the old-fashioned horse and carriage! Many of the shorter journeys could be made on foot or by bicycle.

Many cities are now trying to discourage people from using cars. Historic city centres with their narrow roads are not well suited to cars and lorries. Cars contribute to air pollution (Figure 8) and are more likely to be involved in accidents than public transport. Then there is the problem of parking in often-crowded cities with little available space.

Activity

8 Study Figure 8.
   a) Why do you think many people prefer to use cars for journeys in cities?
   b) Make a list of some of the problems caused by cars in cities?
   c) Do you think cars should be banned from city centres? Explain your answer.

▲ Figure 8 Air pollution

Throughout Europe, cities are exploring ways to reduce the use of private cars. Many cities have invested heavily in public transport systems, which include metros, trams and buses (Figure 9).

The European Union is promoting **sustainable city** mobility. This involves allowing people and goods to move freely and safely without damaging the environment. Figure 10 describes some transport initiatives that have recently been introduced in European cities.

▲ Figure 9 Transport – trams in Tallinn, Estonia, buses in Dublin, Ireland

Nicosia, Cyprus
There are bus lanes, more bus stops and buses have priority at traffic lights.

Prague, Czech Republic
Heavy lorries are not permitted in the city centre.

Tallinn, Estonia
Electronic displays available on buses and at stops to improve information for the public.

Amsterdam, Netherlands
'Park and Bike' schemes have been introduced and the many waterways in the city are used by water taxis.

Lisbon, Portugal
A new smart-card ticketing system called 'Lisboa Viva' enables people to access tickets more easily and to make use of the city's metro and buses.

Lille, France
Biogas is used as the main fuel for the city's buses.

Rome, Italy
Electronic gates are used to monitor and charge non-registered vehicles entering a Limited Traffic Zone in the historic city centre.

zona traffico limitato

London, England
Congestion charge for vehicles in the city centre has been introduced to encourage the use of public transport.

Ljubljana, Slovenia
Investing in sustainable biofuel technology for buses and taxis.

Bucharest, Romania
New trolley-buses and trams have been introduced.

Athens, Greece
A tram network was constructed in time for the Olympics in 2004. It is reliable and does not emit pollutants.

Vilnius, Lithuania
New cycleways and cycle parks have been used to encourage school children to cycle to school.

Stockholm, Sweden
Since 1996 biogas from sewage has been used to power many of its buses.

Valetta, Malta
In 2006 a free Park and Ride system started in the city. Free bicycles are provided as an alternative to buses.

▲ Figure 10 European transport initiatives

Activities

9 Study Figure 10. For this activity you will need to refer to the atlas map in the inside back cover. You will also need a blank outline of Europe.

a) Locate the cities listed in Figure 10 onto your outline map.

b) Write the name of each city and write a short label describing what has been done to address the problems of traffic.

c) Which city makes use of its canals and waterways to provide transport for people?

d) Why did Athens construct a new tram system?

e) Why do you think Prague has banned heavy lorries from the city centre?

f) Several European cities have introduced vehicles that use alternative fuels. Do you think this is a good idea? Explain your answer.

g) Cycle lanes have been introduced in many cities throughout Europe. Why do you think they are such a popular idea?

h) In France, some cities have free bicycles available at train and bus stations for people to use. What do you think of this idea?

10 Study Figure 6 (page 83), which shows the centre of Tallinn, Estonia.

a) Notice that many of the streets in the centre of Tallinn have small blue arrows drawn alongside. What is the meaning of these arrows?

b) Why is this form of traffic management often used in city centres?

c) Why do you think there is a ring road (shown yellow on the map) around the historic centre of the city?

d) What are the three types of public transport available in the city (apart from the mainline trains)?

e) The city authority is trying to encourage people to use public transport rather than private cars. Do you think this is a good idea? Explain your answer.

RESEARCH

Figure 10 contains several good ideas for reducing the problem of transport in cities. Conduct a study of your own local town or city to find out how it is coping with transport issues.

- Suggest the problems and issues associated with transport in your local town or city (e.g. congestion, pollution, too many cars, etc).

- What is currently being done to address these issues?

- What else do you think could be done? Make some detailed suggestions, using annotated maps and diagrams to illustrate your proposals.

ICT ACTIVITY

Study transport developments in a European city of your choice. Start by accessing the case studies on the European Union's website at http://ec.europa.eu/transport

Scroll through the examples and select a city that interests you.

Conduct some additional research using a Google search.

Write a short report on your chosen city, using photos if possible. Comment on whether you think the schemes are good ones and why.

D European Green Cities

In 1996, the European Union set up the European Green Cities Network to encourage cities to develop sustainable building projects. A huge range of sustainable projects has taken place across Europe, involving housing, transport and the redevelopment of old buildings.

Most of the Green Cities projects are small scale and they all involve local communities in their planning and design. This is a different approach from planning in the past, where new developments were often imposed on people by local authorities. Such developments were often unpopular with the local communities and, as a result, were often badly looked after.

Let's have a look at two of the many projects promoted by the Green Cities Network.

1 Low-energy houses in Radstadt, Austria

In the town of Radstadt, just south of Salzburg in Austria (see the atlas map in the inside back cover), 36 solar-powered low-energy houses have been built (Figure 11). The domestic hot water is solar heated, rainwater is collected and reused, and the air ventilation works with heat recovery. The walls, roofs and windows are all insulated to conserve energy and retain heat. Attractive green areas surround the houses and the quality of life of the residents is high.

2 Building renovation in Volos, Greece

In the Greek town of Volos an old grain sanitation building, used for disinfecting crops, was closed down and abandoned in the 1970s (Figure 12). It has now been modernised and is currently the Regional Energy Centre, promoting energy efficiency and the use of alternative energy sources (Figure 13). The building is well insulated, has natural ventilation and is powered by solar energy.

▲ **Figure 11 Low-energy housing, Radstadt, Austria**

Activity

11 Study Figure 11. One of the housing units in the photograph is for sale. You work for a local estate agent and have been asked to write a short feature for the local newspaper to advertise the property.

a) Look closely at the photograph and re-read the text to identify the key selling points of the property. Make a list in rough.

b) Now write your feature for the local newspaper. You can use your imagination to describe what the house is like inside! The aim of your feature is to get people to contact the estate agent to ask for further information.

▲ **Figure 12   Grain sanitation building (old)**

Activity

12 Study Figures 12 and 13.

a) What changes have been made to the original building?

b) Do you think the renovation of the building has been a success? Explain your answer.

c) Is there anything else that could be done to improve the building further?

d) Do you think it is a good idea to retain old buildings and renovate them rather than demolishing them? Explain your answer.

RESEARCH

Identify a building or a housing area in your local town or city that you think could be improved in a sustainable way. Visit the area to take some photographs. Try to talk to some local people about how the building or housing area could be improved. What are the needs of the local community?

Make some suggestions (using annotated photographs) to show the improvements you would make. Give reasons for the improvements that you suggest. You will find a good deal of useful information about energy saving on the internet. For example, the BBC have an interesting site at www.bbc.co.uk/homes/housekeeping/wastenot_index.shtml

ICT ACTIVITY

Conduct your own research into one of the Green Cities Network projects at www.europeangreencities.com/demoprojects/demoprojects.asp. Scroll through the various projects and find one that you are interested in. Produce your own short report on the project. Include photographs in your report. Say why you think your chosen project is a good idea.

▲ **Figure 13   Grain sanitation building (renovated)**

E Issue: Is the 'Mose' project the best way to protect Venice from flooding?

Many people consider Venice in Italy to be one of Europe's finest cities. With its network of canals and characteristic gondola boats, its magnificent churches and fine art galleries, Venice has drawn visitors from across the world for hundreds of years (Figure 14).

However, built on a number of islands in the middle of a lagoon (Figure 15), Venice is constantly under threat from the sea and flooding is becoming more frequent. On average, flooding now occurs on 200 days a year. In the early 20th century, it only happened on seven days a year. People have been forced to abandon living on the ground floor and have even filled in ground-floor windows (Figure 16). Whilst the flooding is partly due to rising sea levels, the extraction of groundwater and methane gas for local industries has caused the city to sink by as much as 30 cm in the last 100 years.

In 2008, work started on an ambitious and very expensive project to construct a line of dams across the mouth of the lagoon to prevent flooding by the sea (Figure 17). Whilst many of the residents of Venice are relieved that the work

has begun, a number of concerns have been raised:

- How will the lagoon wetland be affected if dams are built? Will wildlife be affected?

- Will the dams cope with the projected rise in sea levels, due to global warming? Research suggests that the dams will only be effective for about 100 years.

- Will the dams simply create a vast pond, making flooding far more of a problem?

- How will the natural coastal systems – the currents and the movement of mud and sand – be affected by the dams?

- At a cost of 4.3 billion euros, are the dams too expensive? Is it a sustainable option?

- Should the people of Venice just learn to live with more frequent flooding?

The news extract in Figure 18 on page 92 outlines some of the views about the project.

▲ **Figure 14 Venice**

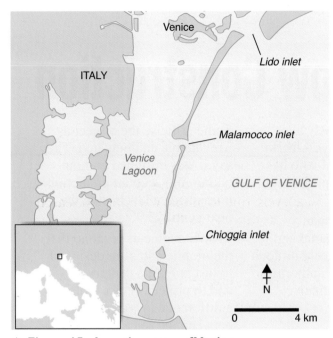

▲ Figure 15   Location map of Venice

▶ Figure 16 Venetian house with filled-in windows

The project, called 'Mose' (Italian for 'Moses'), is due to be completed in 2012 at a cost of 4.5bn euros (£2.9bn).

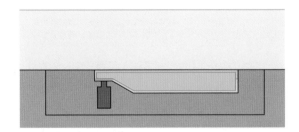

It involves a series of 78 interconnected dams that normally lie flat on the sea bed.

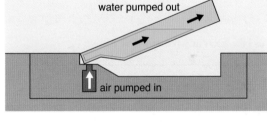

When flooding is likely to occur, the dams are filled with compressed air, forcing the water out and causing the gates to rise on hinges. They form a slanting barrier to the rising tide.

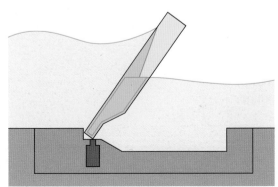

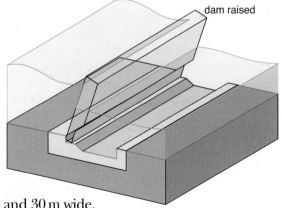

The gates are about 5 m high and 30 m wide.

▲ Figure 17   How will the 'Mose' project protect Venice from flooding?

Protests Fail to Slow Construction

Project supervisor, Claudio Mantovan, says the project is on schedule. Some 37 per cent of the work has been completed, and MOSE should open as planned in 2012. One key element already finished is a navigation lock, to allow large ships to enter the lagoon when the gates are up.

Mantovan says a few days of work have been lost due to peaceful protests by environmentalists and others.

"In order to build trenches for the MOSE gates, they are going to dig up millions of cubic metres of sea bed and replace it with cement, which could seriously alter the ecosystem," says Alberto Vitucci, a journalist who has been covering the project for years.

"The entire mechanism will be underwater, making maintenance extremely difficult and costly. And the authorities never took any alternative projects into serious consideration."

Other proposals to control flooding in Venice have included narrowing the inlet channels to reduce the water flow from the sea into the lagoon, and banning tankers and large ships from entering.

Some criticise the project as irreversible and outdated. They say it was designed without taking into account predictions on rising sea levels over the next century.

MOSE engineers respond that the mobile gates are designed to last at least a century, and to protect Venice from a difference in water level between the sea and lagoon of up to two metres.

The latest prediction, of the United Nations Intergovernmental Panel on Climate Change, is for a 30- to 60-cm increase by the end of this century.

▲ Figure 18   Extract from **NPR News**

Activities

13 Study the information on pages 90-92 and answer the following questions.

a) Why does Venice suffer from flooding?

b) What is the evidence that the problem is getting worse?

c) Describe, with the aid of a diagram, how the 'Mose' project is designed to stop future flooding.

d) Make a list of the objections to the 'Mose' project.

14 Having completed question 13, you should now have a good understanding of this controversial project. Imagine that you lived in the house shown in Figure 16. Write a letter to the local newspaper putting your point of view. Should the work on the project be stopped as the protesters want, or should it continue?

Farming in Europe

CHAPTER

7

In this chapter you will study:

- patterns of farming in Europe
- farming on a mixed farm in the UK
- reindeer herding in the sub-Arctic
- orange production in Valencia, Spain
- the issue of chicken welfare and production of cheap supermarket chickens.

A Farming patterns in Europe

Look at Figure 1. It shows the main types of farming in Europe. Notice that there is a great variety of farming, from forestry and reindeer herding in the north to cereals in the centre and dairying in the west. Towards the south of Europe vineyards and orchards are common. In this chapter we are going to study three very different types of farming in Europe: a mixed farm in the UK, reindeer herding in Norway and oranges in Spain.

On a continental scale the main factor affecting farming is the climate. The wet mild climate in Western Europe suits dairying. The drier climate in central Europe favours crops such as wheat and barley. The colder climates of the north are not good for farming, which is why forests are common. In the south around the coast of the Mediterranean the hot and sunny climate is ideal for grapes, citrus fruit and tomatoes.

Figure 1 is a simplified map of farming. In reality, within each type of farming there is considerable variety due to local conditions of climate, soils and markets. Look at Figure 3, which shows in more detail the farming types on the Italian island of Sicily.

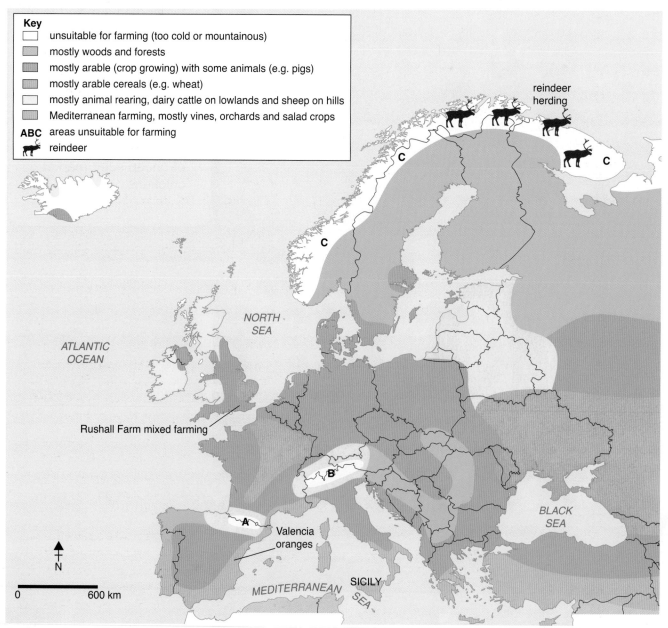

Key
- unsuitable for farming (too cold or mountainous)
- mostly woods and forests
- mostly arable (crop growing) with some animals (e.g. pigs)
- mostly arable cereals (e.g. wheat)
- mostly animal rearing, dairy cattle on lowlands and sheep on hills
- Mediterranean farming, mostly vines, orchards and salad crops
- **ABC** areas unsuitable for farming
- reindeer

▲ Figure 1 Farming patterns in Europe

Pig farming in
Demark

Vineyards in Italy

Cereal farming in
France

▲ Figure 2

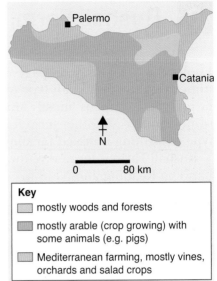

■ Palermo

■ Catania

N

0   80 km

Key

☐ mostly woods and forests

☐ mostly arable (crop growing) with
some animals (e.g. pigs)

☐ Mediterranean farming, mostly vines,
orchards and salad crops

▲ Figure 3   Farm types in Sicily

Activity

1   Study Figure 1. For this activity you will need a
blank outline map of Europe. You may also need
to refer to the atlas maps in the inside front and
back covers.

a)   Make a copy of the different farming patterns
in Figure 1 onto your outline map.

b)   Use colours to show the different types of
farming and explain their meaning in a key.

c)   What is the main type of farming in the
Netherlands?

d)   What is the main type of farming in Estonia?

e)   Why do you think most of Iceland is
described as being 'unsuitable for farming'?

f)   Name the three other parts of Europe
labelled **A B C** that are described as
'unsuitable for farming'?

g)   Try to explain why these areas are not
farmed.

h)   Write a few sentences describing how
climate affects the pattern of farming in
Europe.

ICT ACTIVITY

Figure 3 shows the pattern of farming in Sicily, Italy.
The aim of this activity is to try to explain the
variety of farming on the island. Why are vineyards
and orchards found near the coast? Why does the
centre of the island have arable farming with grazing
and woods? See if you can discover if there are
variations in climate, soils or relief.

A good website to start with is
www.viaggia.com/map/index.html Here you will find
a map and a description of the land use and the
climate. See what else you can discover as a
Geography detective!

RESEARCH

Carry out some research to find out where
fresh fruit and vegetables have come from. To do
this you need to visit a local supermarket, either
as part of a school trip, or with your family.

Write out a list of fresh fruit and vegetables
available in the shop. Alongside, write down the
country of origin – you should find this displayed
with the price. If no country of origin is given, it
probably comes from the UK.

Combine your results in class. Make a separate list
of the fruit and vegetables that have come from
elsewhere in Europe. Are there any patterns? Do
most products come from southern Europe?
Suggest reasons for the origins you have identified.

B Rushall Farm, a mixed farm in the UK

Look back to Figure 1 on page 94 and notice that the dominant type of farming in Europe is 'mostly arable with some animals'. Another term for this type of farming is **mixed farming**.

Rushall Farm is located midway between Reading and Newbury in Berkshire (Figure 1). It is a good example of a mixed farm. The farmer, Mr Bishop, grows a range of cereal crops, looks after cattle and sheep, and manages natural grassland and woods (Figure 4). Since 2000, the farm has been run as an **organic** farm. This means that the farmer does not use any artificial chemicals or fertilisers.

Throughout the year, Mr Bishop looks after his large flock of nearly 850 sheep and herd of 80 cattle. The sheep spend most of the year grazing outside. They come into the barn for lambing in December, where they are fed beans, oats, wheat, hay (dried grass) and silage (fermented grass). Shearing takes place in May and June (Figure 5). The cattle graze outside in the summer, but are moved inside in October for the winter (Figure 6).

In addition to growing crops to feed the animals, Mr Bishop grows barley for malting and wheat for making into bread. Sowing takes place in February or September, and harvesting is in July and August.

Farming today is not just about growing crops and keeping animals. Farmers have an important role in looking after the countryside. Mr Bishop is keen to maintain the natural environments on his farm. He looks after an area of woodland and manages a wildflower meadow (Figure 4). Grants are available from the European Union to help him manage these wildlife habitats that are rich in birds, wildflowers and animals (such as deer, hares and badgers).

▲ **Figure 4 Wildflower meadow**

▲ **Figure 5 Sheep shearing**

Activities

2 Study Figure 6. It shows some details of the farming year at Rushall Farm.

a) Make a large copy of Figure 6.

b) Use the text on page 96 to add additional labels describing the activities on the farm. Position the activities in their correct places alongside the line.

c) During which months of the year do you think the farmer is busiest? Explain your answer.

d) Why do you think the farmer carries out most of the maintenance tasks in the winter?

e) Why do you think the cattle are kept inside during the winter?

f) Do you think the farmer is wise to grow animal feed on his farm? Why?

g) What time of year do you think it would be best for Mr Bishop to take his main holiday? Give reasons for your answer.

3 Study Figure 7. It lists a number of features on Rushall Farm. In Book 1 you learned that farms can be described as **systems** with inputs, processes and outputs.

- **Inputs** – factors such as the climate, buildings, seeds and land that affect what a farmer does

- **Processes** – actions that take place on the farm such as ploughing

- **Outputs** – products such as meat, milk and straw.

a) Draw a large rectangular box and use a ruler to draw vertical lines to divide it into three.

b) Work in pairs to sort the features listed in Figure 7 into the three boxes, one for inputs, one for processes and one for outputs. Write these as headings.

c) Add some simple sketches or use colour for the writing to make your diagram attractive.

d) Write the title 'Rushall Farm System' at the top of your diagram.

4 Study Figure 4.

a) Describe the scene in the photograph. What are the natural habitats that you can see? Can you name the red flowers?

b) Do you think it is important to keep meadows and hedges on farms? Explain your answer.

c) Do you think farmers should get paid to look after the countryside? Why?

d) Schools are encouraged to visit the farm to learn about farming and countryside management. Do you think this is a good idea? Why?

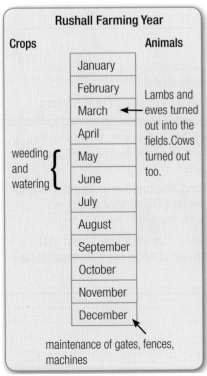

▲ **Figure 6** Farmer's Year

▲ **Figure 7** Features of Rushall Farm

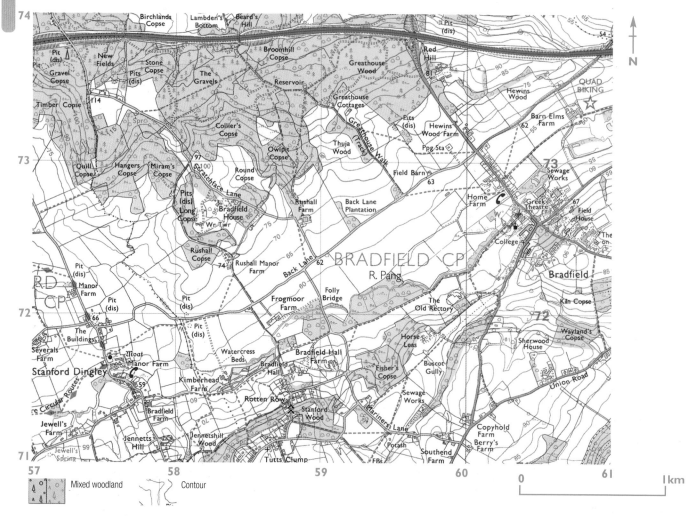

Mixed woodland   Contour

▲ **Figure 8   OS map (1:25,000) of Rushall Farm**

Activity

5   Study Figure 8. It is a 1:25,000 OS map showing the location of Rushall Farm.

a) What is the six-figure grid reference of Rushall Farm?

b) Locate nearby Rushall Manor Farm. This is where school groups meet to visit the farm. What is its six-figure grid reference?

c) If you walked across the field from Rushall Manor Farm to Rushall Farm would you be going uphill or downhill? Explain your answer.

d) Rushall Farm is on the north side of the valley of which river?

e) The nearest large village is in grid square 6072. What is its name?

f) What is the main land use in grid square 5873?

g) Grass needs plenty of water. Where would you expect to find grass fields in grid square 5972 and why?

h) The land around Rushall Farm faces towards the south. Can you think how this is an advantage to Mr Bishop's cereal crops?

i) Mr Bishop's land has very few steep slopes. How is this an advantage for mixed farming?

ICT ACTIVITY

Find out more about the activities at Rushall Farm by accessing the Farming and Countryside Education (FACE) website at www.face-online.org.uk . Click Farm Profiles and then click Rushall Farm.

- Imagine that your school is interested in visiting Rushall Farm. Find out what pupils would be able to do on the farm. Which activities would you be interested to do and why?
- Click 'Land Use' and scroll down to find out more about the management of the grasslands and the woodland. How is the government supporting the farmer in managing these natural environments?
- Click 'Wildlife'. What wildlife is found on the farm? If the woods, hedges and natural meadows were to be removed, what effect do you think this would have on the range of wildlife found on the farm?
- Choose to research one of the following activities that take place on the farm:
  - Sheep   • Cattle   • Crops

C Reindeer herding in Norway

Turn back to Figure 1 on page 94. Look for the location of reindeer herding in Europe. Notice that it is found in the far north of Norway and Finland, at the very edge of Europe. With their thick hides and broad feet, reindeer are well suited to living in this cold environment where snow lies on the ground for several months in the winter. Reindeer are primarily kept for meat. Their hides can also be used for making clothes and handicrafts.

Reindeer herding is a traditional type of farming carried out by the Sami people who have lived in this remote and hostile part of Europe for hundreds of years. Johan Henrik is a reindeer herder (Figure 9). Notice the colourful clothes that he is wearing. These are the traditional Sami clothes often made by women during the long dark winters.

Johan is skilled in looking after his large herd of reindeer and in helping them to survive the harsh conditions. Each spring, he leads his herd on a migration to find summer pasture (Figure 10, page 100). Look at Figure 11, page 100, which describes the reindeer-herding year.

Activities

6 Study Figure 9 on page 100.

   a) Describe the traditional clothes worn by Johan Henrik.

   b) Why do you think the traditional Sami clothes are so brightly coloured?

   c) How can you tell from his clothes that Henrik lives in a cold climate?

   d) How are reindeer well adapted to live in this hostile environment?

   e) Johan is harnessing a sled to be used on the annual migration. What do you think the sled will be used for? (See Figure 11 on page 100 for some help!)

7 Study Figure 10 on page 100.

   a) Make an estimate for the number of reindeer in the herd.

   b) How many people can you see in the photograph?

   c) Describe the shape of the reindeer herd.

   d) What is the landscape like?

   e) In what season do you think the photograph was taken? Use the information in Figure 11 to help you answer this question.

   f) Imagine that a reindeer wandered off. Do you think it would be easy to find the missing reindeer? Explain your answer.

◀ **Figure 9   Johan Henrik with reindeer**

Activity

8 Study Figures 11 and 12.

a) Make a large copy of Figure 12 to describe the reindeer-herding year. Notice that the diagram is in the form of a cycle.

b) Select some interesting facts to write in each box and draw some simple and colourful sketches alongside to show what happens in each season.

c) Give your diagram a title.

▲ Figure 10 Reindeer migration

Spring

The spring migration takes the reindeer from their inland wintering grounds to the coast. Here the warmer conditions provide plenty of grass for grazing. Salt is also a valuable part of their diet. In the spring, calving takes place.

Summer

During the summer, the herds are found along the coast or on small islands, which they swim to. In this season, the adult reindeer and their calves have to eat as much food as they can to help them survive through the winter. The herders live in temporary tents, keeping an eye on their animals. Some animals are slaughtered during the summer for their hides and for meat.

Autumn

In September and October, the herds turn inland, feeding on grasses and mushrooms as they go. The reindeer have been used to the freedom of the summer pastures and can be difficult to keep together. A few wander off and get lost. The autumn is when some animals are sold for meat. The annual 'rut' (mating season) takes place during this time. The males (bulls) fight to establish a 'pecking order takes place.

Winter

Herds are now split into smaller herds. This is because the animals compact the snow, making it hard to find the lichens on which they depend at this time of year. Smaller herds cause less trampling of the overlying snow. Reindeer naturally stay close together in the winter. The herders spend time with their families during the winter, sitting out the two months of total darkness. A few reindeer are slaughtered for meat.

▲ Figure 11 The reindeer-herding year

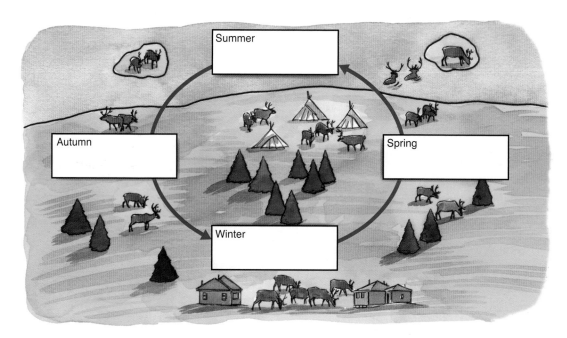

Summer

Autumn

Spring

Winter

▲ Figure 12   Reindeer herding cycle

D Oranges in Valencia, Spain

When did you last eat an orange or drink some orange juice? Oranges are one of the most common and useful types of fruit available to us. Apart from fresh fruit and juice, oranges can be used to make jams and marmalade. The white orange blossom can be dried and used to make tea, and bees feeding on the blossom make particularly delicious orange-flavoured honey. Orange peel can be crushed to produce oil used in perfumes and for aromatherapy. Peel can also be used in the garden as a slug repellent! Oranges are, of course, extremely good for us, as they contain vitamin C.

Oranges grow on trees usually grouped together to form orchards or groves (Figure 13). They are grown throughout the world in warm and sunny climates (Figure 14 on page 102). In Europe, oranges are grown in countries bordering the Mediterranean (such as Spain, France, Italy and Greece).

One of the main orange-growing areas in Europe is Valencia in Spain (Figure 1, page 94). Some people call this region the 'Orchard of Spain'. Valencia is well suited to growing oranges. The high temperatures and long hours of sunshine help the oranges to grow and ripen,

and there is enough moisture during the year to enable them to form juicy fruit. The mild winters mean that damaging frosts are rare too.

The oranges in this region are a variety called 'Valencia oranges'. They are late fruiting (May to July) oranges and are particularly sweet. Most of the oranges are used to make orange juice before being exported across Europe and the world.

▲ Figure 13   Orange groves, Valencia, Spain

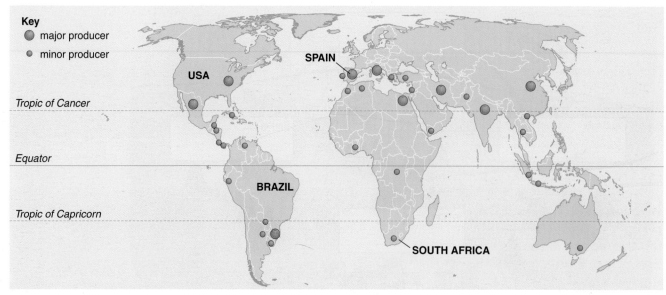

Key
● major producer
• minor producer

Tropic of Cancer

Equator

Tropic of Capricorn

USA

SPAIN

BRAZIL

SOUTH AFRICA

▲ Figure 14 Map showing orange producers worldwide

Activities

9 Study Figure 14. It shows where oranges are grown around the world.

a) Apart from Spain, which European country is a major producer of oranges?

b) Which South American country is a major producer of oranges?

c) Locate the USA on Figure 14. Are oranges grown in the east or the west of the country?

d) 'The major producers of oranges form a broad belt running west-east across the world.' Do you agree or disagree with this statement?

e) Why do you think oranges are not grown in the UK?

f) Notice that oranges are grown in South Africa. If you were to visit this part of South Africa, what do you think the climate would be like?

g) How good is your knowledge of world geography? Can you name three other countries where oranges are grown?

10 Study Figure 15.

a) Make a copy of Figure 15.

b) Add the following labels to describe how orange trees are well adapted to the hot and dry Mediterranean climate.

● Thick bark that is resistant to summer fires

● Oranges have thick waxy skins to conserve water

● Leaves shade the fruit and help to keep the soil beneath the tree moist

● Extensive roots search for water in the soil.

c) Now write some of the uses of oranges (see page 101) in the form of boxed labels positioned around your sketch of the orange tree. Perhaps you should write them in orange!

▲ Figure 15 Orange tree

E Issue: Should chickens suffer to provide us with cheap food?

Did you know that around 855 million chickens are reared for their meat in the UK every year? This is equivalent to 14 chickens for every single person living in the UK!

About 95 per cent of these chickens are kept indoors, packed densely into vast sheds (Figure 16). With less space than an A4 sheet of paper each and fed high-protein food, some chickens live for only 40 days before ending up on the supermarket shelf selling at just £2.50. Many chickens simply collapse onto the ground, as their thin legs are unable to support their huge over-fed bodies.

So, why is this happening in a country often mocked for being soft on animals? One of the reasons is the power of the supermarkets in driving down prices for farmed produce. In order to be cost-effective, chicken farmers have responded by producing more and more chickens. This means cramming them into sheds and fattening them up as quickly as possible. Inevitably, this has led to concerns over the welfare of the birds.

Kept in the open and allowed to range freely, chickens like to forage on the ground for food (Figure 17). They are inquisitive and respond to their surroundings. These freedoms do not exist for millions of birds reared intensively in dark and overcrowded sheds, unable to flap their wings or move around.

In 2008, the RSPCA launched a campaign calling on supermarkets to stop selling mass-produced standard chickens. Celebrity chefs, like Hugh Fearnley-Whittingstall, presented TV documentaries highlighting the poor conditions on many chicken farms. They urged the public to buy free-range or organic chickens.

▲ Figure 16 Chickens crammed into a barn to be fattened up

▲ Figure 17 Free-range chickens

Activities

11 Work in pairs or small groups to discuss the following questions:

a) Why do supermarkets sell chickens at very low prices?

b) Should shoppers buy cheap mass-produced chickens?

c) Why do farmers rear chickens in overcrowded conditions?

d) Who is to blame for the poor welfare of chickens: supermarkets, shoppers or farmers?

12 Study the RSPCA's guidelines in Figure 19. Chickens reared in the conditions laid down by the new standard will be labelled 'Freedom Food'. Shoppers are being encouraged to buy only chickens with this label.

a) Work individually or in pairs to design a campaign leaflet to be distributed outside supermarkets informing customers of the RSPCA's campaign. In your leaflet you need to explain why the RSPCA has introduced the new Freedom Food Label and how it helps to improve standards of chicken welfare. Take a look at the animation at (www.supportchickennow.co.uk/freedomfood/index.html). Your leaflet needs to have a clear message so think carefully what it should be. You can design your leaflet either by hand or using ICT (such as Publisher).

b) Suggest where you think the leaflet should be distributed in your local town. Can you suggest other ways of getting your message across to the public?

▲ Figure 18 The RSPCA's Freedom Food Campaign

1 Birds must be given bright lighting during the day and a longer dark period at night to allow them a proper rest period.

2 Birds must be given space to allow them to flap their wings and walk around.

3 Birds must be given an interesting living space including perches, straw bales and pecking objects to encourage activity and the expression of natural behaviour.

4 Birds must be selected from genetically slower-growing breeds, to help overcome the serious welfare problems associated with faster growth rates.

▲ Figure 19 RSPCA welfare standards (labelled 'Freedom Food') for meat chickens

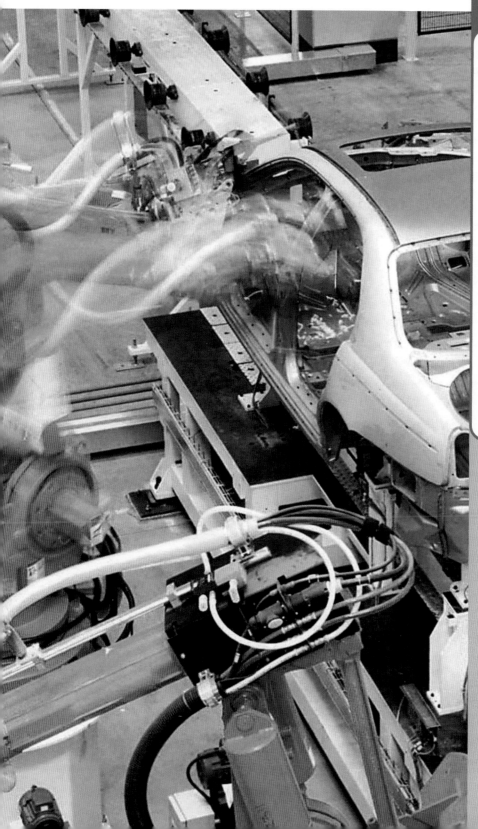

Energy and Industry

A Patterns of energy in Europe

We consume a great deal of energy in Europe. We demand energy to power our industry, provide electricity for lighting and heating and to fuel our cars, lorries and aeroplanes. Can you think of other demands for energy in Europe?

Look at Figure 2. Notice in the key that there is a range of energy sources in Europe. Some are **non-renewable** fossil fuels, such as coal, oil and gas (these resources will eventually run out or simply become too expensive to exploit). Other sources of energy are **renewable** (these forms of energy will not run out). Hydroelectric power, which uses the power of running water, is an important type of renewable energy in Europe.

Today, most of the energy that we use is in the form of electricity. Both non-renewable and renewable sources of energy are used to generate electricity in power stations (Figure 1).

▲ Figure I Electricity power station, Germany

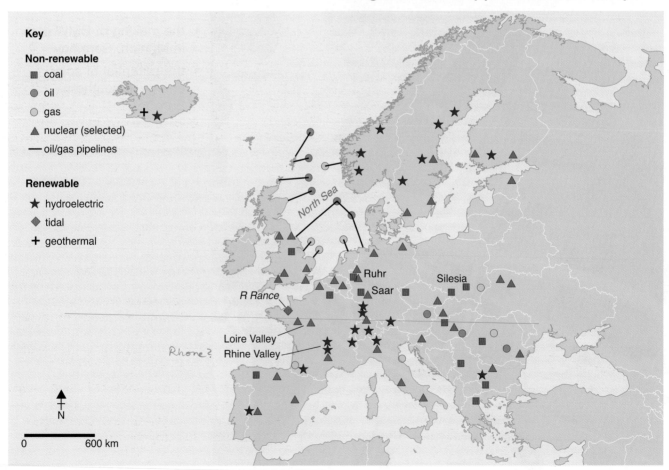

▲ Figure 2 European energy sources

In the past, coal was the most important source of energy in Europe. It fuelled the Industrial Revolution in countries such as the UK and led to the growth of towns and cities. Much of the current wealth of Europe has resulted from the growth of industry originally based on the coalfields.

More recently, oil and gas have become major sources of energy. These are cleaner fuels and easier to transport by pipeline. Oil has a great many uses, both as a source of energy and also as a raw material in the chemical industry. Improved technology has enabled reserves of oil and gas to be exploited in hostile environments such as the North Sea (Figure 3).

▲ **Figure 3   North Sea oil platform**

Activity

1   Study Figure 2. For this activity you will need to refer to the atlas maps on the inside front and back covers.

a)   Which **one** of the following countries does not have coal reserves: UK, Germany, Ireland, Poland?

b)   Which **one** of the following countries has geothermal energy: France, Poland, Portugal, Iceland?

c)   Which **one** of the following countries does not have hydroelectric power: Norway, Sweden, Finland, Romania?

d)   What is the energy source on the Spain/Portugal border?

e)   Which two energy sources are exploited in the North Sea?

f)   How is the energy in the North Sea transported to land?

g)   Name one mountain range in Europe (other than the Alps) where hydroelectric power is produced.

h)   Why are hydroelectric power stations found in mountainous areas?

i)   What type of energy is produced in the Loire Valley in France?

j)   In which country is Silesia and what is the main source of energy there?

▲ **Figure 4   River Rance tidal power scheme**

Nuclear power is widespread in Europe (Figure 2, page 106). It is regarded as a clean and efficient form of energy and it does not pollute the atmosphere. However, in using uranium as a fuel it is a non-renewable form of energy.

At present, hydroelectric power (HEP) is one of the most important and widespread forms of renewable energy in Europe (Figure 2). It uses the power of running water to turn turbines and generate electricity. This explains why HEP plants are usually found in the mountains, where there is high rainfall and fast-flowing rivers.

The European Union is encouraging countries to increase their renewable sources of energy. In addition to hydroelectric power, several other sources of renewable energy are being used in Europe. In northern France, there is a tidal power station on the River Rance (Figure 4). This uses the power of the rising and falling tides to generate electricity. Elsewhere there are wind farms in Denmark, geothermal power stations in Iceland, biomass (fuel from vegetation) in Sweden (Figure 5) and solar power plants in Spain.

▲ **Figure 5   Biomass power station in Sweden, using woodchip and other wood waste**

Activities

2 Study Figure 6.

a) On what single source of energy does Poland depend for generating electricity? Give the percentage in your answer.

b) What percentage of Poland's electricity is generated by renewable sources?

c) What is Poland's most important renewable energy source?

d) Look at Figure 2 on page 106. Apart from coal, what is the other main energy source in Poland?

e) What percentage of Poland's electricity is generated by this source?

f) Can you suggest some disadvantages with Poland's current energy balance? (How sustainable is it? Is it good for the environment?)

g) What do you think should be done in the future to make Poland's energy supply more sustainable?

3 Study Figure 7.

a) Use the data in Figure 7 to draw a pie chart similar to the one in Figure 6. Remember that to convert percentages to degrees you need to multiply by 3.6.

b) Use colours to show each segment and explain the colours in a key.

c) Write a title.

d) Write a few sentences comparing Finland's pie chart to that of Poland.

e) Do you think Finland has a more sustainable future than Poland? Explain your answer.

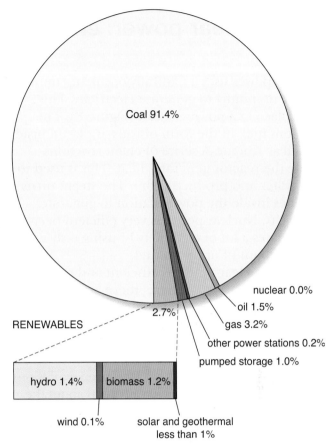

▲ Figure 6   Electricity production, Poland

Renewable sources (33%)		Non-renewable sources (67%)	
biomass	14%	nuclear	33%
hydroelectric power	19%	gas	17%
		coal	16%
		oil	1%

▲ Figure 7 Electricity generation by fuel in Finland (2005)

ICT ACTIVITIES

- Find out more about the tidal power station on the River Rance in France. What is a tidal barrage and why does one have to be constructed? How do the tides generate electricity? Try to find a diagram that describes how it works. Find a photograph to illustrate your research.
http://home.clara.net/darvill/altenerg/tidal.htm
www.reuk.co.uk/La-Rance-Tidal-Power-Plant.htm
http://en.wikipedia.org/wiki/Rance_tidal_power_plant

- The pie chart in Figure 6 came from an excellent website www.energy.eu/#renewable. Access the website and click 'Factsheets EU member States' on the left-hand side. This will reveal a drop-down menu with each country listed. Click your country of choice and scroll down to find data. Select a country of your choice and use the statistics to draw your own pie chart. Write a few sentences describing the balance of energy sources in your chosen country. How sustainable is it for the future?

B Nuclear power: energy for the future?

Nuclear power uses a naturally occurring mineral (called uranium) to generate electricity. This takes place in a power station (Figure 8). The uranium fuel, in the form of rods, is placed inside a nuclear reactor. A series of chain reactions inside the reactor generates heat. This is used to heat water and produce steam. The steam turns turbines inside the power station to generate electricity. Nuclear power is very efficient because it generates a lot of electricity by using only a very small amount of uranium fuel.

Whilst nuclear power is efficient and non-polluting of the atmosphere, there are some concerns. The waste produced is highly toxic and needs to be disposed of safely. It remains toxic for many hundreds of years and could have serious environmental effects. There is always the threat of a radioactive leak and there have been explosions in the past. Some people are concerned about the links with nuclear weapons too.

▲ Figure 8   Dungeness nuclear power stations, UK

Activities

4 Figure 9 is a flow diagram showing how electricity is generated at a nuclear power station.

a) Make a large copy of Figure 9.

b) Arrange the following statements in their correct order and write them in the empty boxes
- steam turns turbines to generate electricity
- reactions in the nuclear reactor produce heat
- uranium is mined and turned into rods
- electricity is distributed around the country
- water is heated to produce steam.

c) Why is nuclear power an example of a non-renewable type of energy?

5 Study Figure 10. It shows the location of Dungeness nuclear power stations in Kent. Figure 8 is a photo of the power stations. There are two power stations (A and B) on this site, capable of producing enough power to provide electricity for the whole of southeast England.

a) In what grid square are the two power stations situated?

b) In grid square 0716 is the electricity sub-station. You can see the power lines running from it. What is the six-figure grid reference of this building?

c) Why do you think the power stations are on the coast? (Hint; what does a nuclear power station need a lot of?)

d) There is plenty of flat land in this area. Why is this important for the construction of a power station?

e) Notice that there is a rail link to the power stations. Why is this important?

f) What is the meaning of the black dots and circles that cover most of the map extract?

g) Dungeness is a remote location in southeast England. Do you think this was an important factor in locating the first UK power station here in 1965? Explain your answer.

RESEARCH

Carry out an individual project on a renewable energy source from the following list: hydroelectric power, wave, biomass or geothermal. If you haven't done the earlier ICT activity on the River Rance tidal power scheme, you could do tidal power. (Don't do solar energy as you will be studying it later in this chapter.)

- Introduction – describe what it is and why it is renewable

- How it works – describe how the power source is used to generate electricity

- Advantages and disadvantages – try to identify the good and the bad points about your chosen energy source

- Example – include a labelled photograph and location map of a power station.

You will find many sites using a Google search, but a good starting point is http://home.clara.net/darvill/altenerg/nuclear.htm.

▲ Figure 9   Generation of electricity from nuclear power

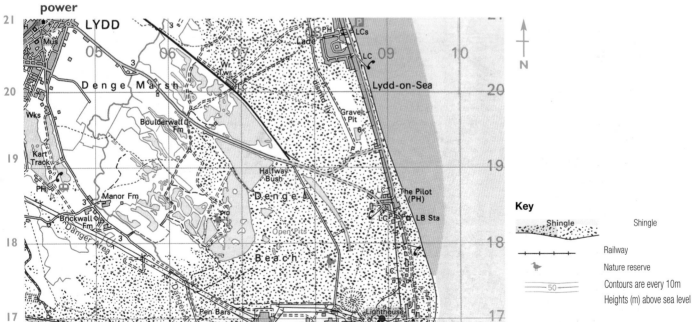

Key

Shingle	Shingle
Railway	
Nature reserve	
—50—	Contours are every 10m
	Heights (m) above sea level

0   1   2km

▲ Figure 10   OS map extract 1:50,000 Dungeness

C From mineral to sweet wrapper: the story of aluminium

Aluminium is one of the most useful metals in the world (Figure 12). You have probably come across it several times today without even realising it. For example, as foil wrapping a bar of chocolate or the container of a fizzy drink. As a strong, yet light and malleable (easily bent and twisted) metal it is widely used in buildings and in making cars and aeroplanes.

Unlike many metals, aluminium is not dug out of the ground as a shiny, silver lump of rock. Instead it has to be made or **manufactured**. This makes it an example of a **secondary industry**.

The raw material used to make aluminium is a naturally occurring rock called **bauxite**. It is usually red/brown in colour and is extracted from huge quarries using giant machines (Figure 11). The extraction of bauxite is a good example of a **primary industry**.

▲ Figure 11   Giant machines mining bauxite

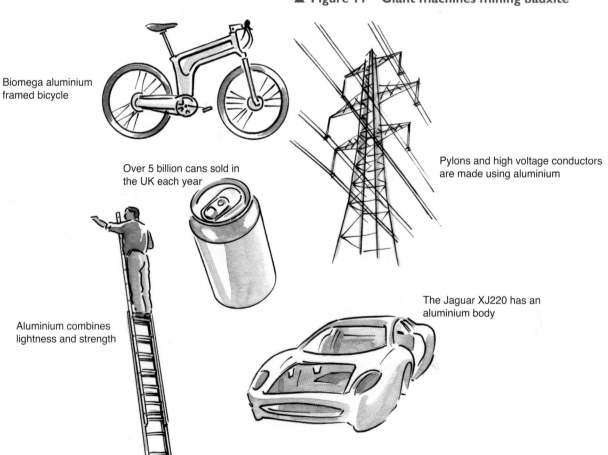

Biomega aluminium framed bicycle

Over 5 billion cans sold in the UK each year

Aluminium combines lightness and strength

Pylons and high voltage conductors are made using aluminium

The Jaguar XJ220 has an aluminium body

▲ Figure 12   Some uses of aluminium

Look at Figure 13 to see how bauxite is turned into aluminium. It takes 4 tonnes of bauxite to make 1 tonne of aluminium. Bauxite is very bulky and expensive to transport. For this reason it is often processed nearby. Alumina is more valuable and is often transported great distances to be turned into aluminium. The process of electrolysis uses huge amounts of electricity, which explains why many aluminium works are located near cheap sources of electricity such as in Iceland (Figure 14).

Bauxite is mined in huge opencast quarries.

Bauxite is crushed and processed into pure aluminium oxide (alumina).

Alumina is converted into aluminium using electricity – this process is called electrolysis.

Aluminium is rolled into sheets or made into wires and cables to be used in other industries.

▲ Figure 13   Flow diagram showing manufacture of aluminium

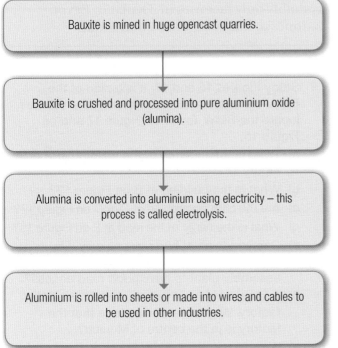

▲ Figure 14   Alcoa aluminium works, Iceland

Activities

6 Study Figure 12.
  a) With the aid of simple sketches, produce a collage showing some of the different uses of aluminium. Include the ones shown in Figure 12 and try to add a few of your own.
  b) For each of the images shown in Figure 12, suggest how aluminium's quality of being lightweight is an advantage.
  c) For which of the images in Figure 12 do you think the quality of strength is particularly important? Why?

7 Study Figures 11 and 13.
  a) Describe the impact of the bauxite mine on the landscape.

  b) Why do you think environmentalists often oppose opencast bauxite mining?
  c) Why is bauxite often processed into alumina close to where it is quarried?
  d) What is electrolysis?

8 Study Figure 14.
  a) Iceland has plenty of cheap electricity available to convert alumina into aluminium. Look back to Figure 1 (page 106) to discover the sources of energy in Iceland.
  b) Describe the main features of the aluminium works in the photo.
  c) The site of the aluminium works in Figure 14 was very carefully chosen. Can you suggest some advantages of the site?

D BMW: a European transnational company

BMW is one of the world's leading producers of cars and other vehicles. The Headquarters of the company is located alongside one of the main car factories in the heart of Munich in Germany (Figure 15).

Built in the 1960s, the Munich car plant produces and assembles the 'Series 3' cars (Figure 16) using up-to-date technology. The factory employs 9000 people (called 'associates') and it produces 200,000 cars a year. There are some 300 separate jobs carried out within the factory. Each car involves the assembly of thousands of separate parts or **components**.

The Series 3 BMW car starts life as a roll of steel that is cut into plates and then shaped into part in the Press Shop (see Figure 15). The different-shaped steel parts are assembled in the Body Shop to produce a car body. Robots weld the parts together. The car body is moved to the Paint Shop, where the body is treated to protect it from corrosion. Four layers of colour are then added to the car body. Elsewhere in the factory, engines are assembled in the Engine Shop (see Figure 15). Finally, the car body is moved to the Assembly Shop, where it travels down a conveyor 3475 m long. Here the chassis, engine and gearbox and interior fittings are installed in the car body, and the various fluids are added so that the car can be driven off the assembly line.

BMW operates all over the world. It is an example of a **transnational company** (TNC). There are 24 sites in 13 countries, with production and assembly sites in Germany, UK, Austria, South Africa, the USA and China, and assembly-only factories in Thailand, Egypt, Indonesia, Malaysia, Russia and India.

Activity

9 Study Figures 15 and 17. It is a plan of the BMW car plant in Munich. To orientate yourself, locate the BMW Tower on Figure 17 and on Figure 15.

a) What is the function of the building at A on Figure 15?

b) What is the building at B on Figure 17?

c) Are the cars assembled at C or D on Figure 17?

d) What is the name of the road at E on Figure 15?

e) Can you think of two reasons why it is important for a car production and assembly plant to have good road access?

f) A large number of people work in the BMW factory. Why is it an advantage that the factory is in the centre of Munich?

g) Why do you think BMW have kept areas of grass and trees around the factory site?

▼ **Figure 15 BMW factory and HQ, Munich**

Paint shop · Body Shop · Engine shop · BMW Tower · Press shop · Assembly shop · A · E

Activity

10 For this activity you will need a blank map outline of the world. Use the atlas map on the inside covers if you need help.

a) Locate the six countries where there are BMW production and assembly works. Shade these countries using a colour of your choice.

b) Now locate the six countries where there are BMW assembly-only factories. Use a different colour to shade these countries.

c) Locate Munich in Germany and label this location as BMW's Headquarters.

d) Explain the colours in a key and give your map a title.

e) BMW is an example of a transnational company. What is a transnational company?

f) Why do you think BMW has production and assembly plants in some of the most-developed countries in the world, such as the USA?

g) Why do you think BMW has assembly-only plants (where the components are assembled not made) in less-developed countries such as Indonesia and India?

▲ **Figure 16   Series 3 BMW car**

▼**Figure 17   Plan of the Munich BMW factory**

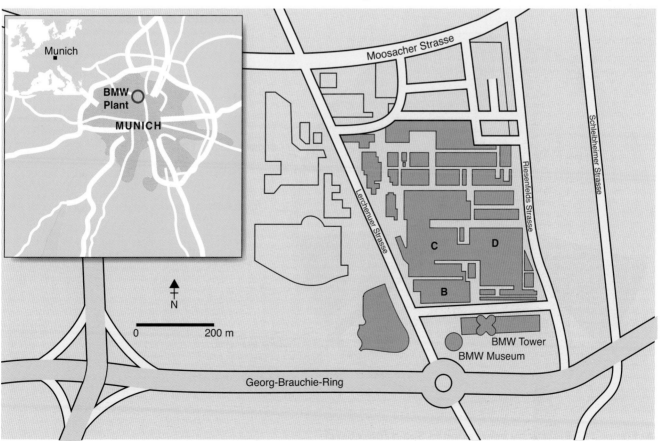

E Issue: Should solar energy be exploited in Southern Europe?

In 2007, the European Union published a map showing the potential for solar energy in Europe (Figure 20). The map shows that the sunny areas of Europe (such as Malta and Southern Spain) have the potential to produce much more solar energy than areas such as Scotland or Northern Scandinavia.

The European Union has set a target of producing 20 per cent of its energy using renewable sources by 2020. By exploiting solar energy in the sunnier parts of Europe, the EU could make a significant step towards its target.

Figure 18 lists the top ten countries currently exploiting solar energy using photovoltaic cells (a small high-tech 'plate' that converts sunlight to electricity). Notice that Germany is way ahead of other European countries, with Spain being the only other major user. Despite being 2nd in the 'top ten', solar energy in Spain only accounts for less than one per cent of its electricity generation (Figure 19). Recognising the potential for solar energy, several new schemes are currently being developed in Spain (Figure 21 on page 118).

Could (should) other Southern European countries make use of solar energy?

Rank	Country	Photovoltaic capacity installed in 2006 (in MWp)
1	Germany	1153
2	Spain	60.5
3	Italy	11.6
4	France	6.4
5	Austria	5
6	United Kingdom	2.8
7	Belgium	2.1
8	Greece	1.3
9	Sweden	0.7
10	Cyprus	0.5

▲ Figure 18 Top ten users of solar radiation in Europe –

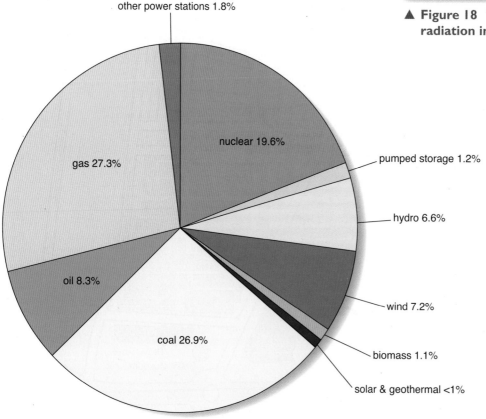

▲ Figure 19 Electricity generation in Spain (pie chart)

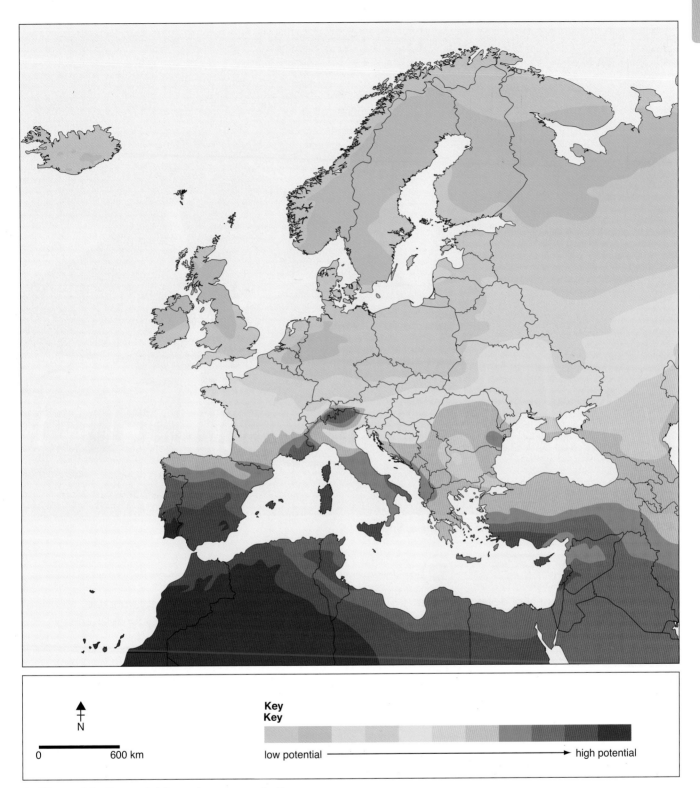

Key
Key

low potential ──────────────────────────────→ high potential

0   600 km

N

▲ Figure 20   Potential for solar energy in Europe

▲ **Figure 21 A solar energy farm in Almeria, Spain**

RESEARCH

Research the potential of developing solar energy for a country of your choice. Your study can be presented electronically, if you wish, or in the form of a printed report.

1 Find out about the generation of electricity using solar energy and photovoltaic cells. This should form an introduction to your study. A good starting point is http://home.clara.net/darvill/altenerg/solar.htm.

2 Use Figure 19 (page 117) to select a country in Southern Europe (other than Spain) that has the potential to develop solar energy.

3 Study and copy a map showing the solar energy potential in your chosen country at http://re.jrc.ec.europa.eu/pvgis/countries/europe.htm. Use an atlas map (or the internet) to help you identify the most favourable parts of your chosen country for solar energy developments.

4 Find out about your chosen country's current energy sources for electricity production by looking at the Country Factsheets at **www.energy.eu/#renewable**. Copy the pie chart for your country and write a few sentences describing the balance of energy sources used to generate electricity.

5 Use a Google search to see whether your chosen country has started to develop solar energy. You might find a project similar to the Almeria project in Spain.

6 Suggest why you think your chosen country should make greater use of solar energy in the future.

CHAPTER

9

In this chapter you will study:

- types and patterns of tourism in Europe
- the advantages and disadvantages of tourism
- Theme Park tourism: Disneyland Resort Paris
- rural tourism in Gozo, Malta
- mass beach tourism developments in Spain.

A Tourism in Europe

Europe is the most important tourism region in the world, both as a destination for tourists (Figure 1) and as a source of tourists. Visitors travel to European countries from all over the world, although most tourists come from other European countries. Europeans themselves travel widely across the world.

Europe has many attractions for tourists (Figure 2). There are the cultural and historic cities of Rome, Tallinn, Paris and London, the many winter sports resorts in the Alps and the beautiful wilderness areas of northern Scotland, Iceland and Scandinavia. The hot and sunny beach resorts along the Mediterranean coast in Spain, France and Italy attract hordes of summer visitors, and people interested in ancient history choose to visit Greece, Cyprus and Italy. Can you think of any other types of attraction to tourists in Europe?

Tourism in Europe has grown rapidly in recent years. People have more money to spend and more holiday time available. Travel has become cheaper and more accessible, particularly since the introduction of cheap 'no frills' flights. Newspapers are full of adverts for cut-price tickets. Look at Figure 3. It shows the European routes available for easyJet flights from just one airport, London Gatwick. With an increasing number of regional airports being used by the budget airlines, new parts of Europe are being opened up for tourism.

▲ Figure 1 Advertisement for Europe

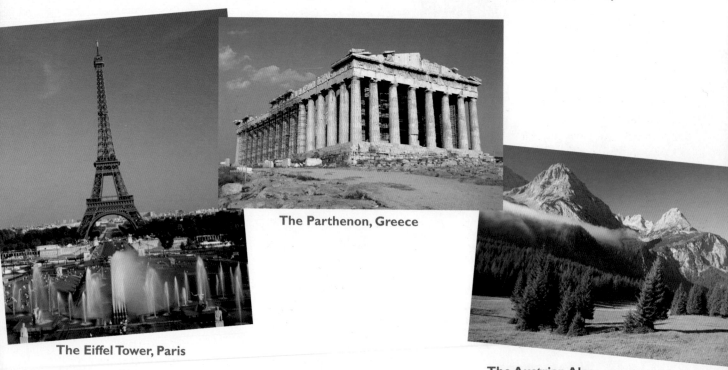

The Parthenon, Greece

The Eiffel Tower, Paris

▲ Figure 2 Tourist attractions, Europe

The Austrian Alps

Activities

1 Study Figure 1. This is a poster produced by Air France in the 1950s.

  a) What are the attractions featured in Figure 1?

  b) What type of people (young, families, older people) do you think the poster is aimed at?

  c) Do you think it is a good poster? Explain your answer.

  d) How do you think the poster could be redrawn for the present day?

2 Study the photos in Figure 2.

  a) For each photo, describe the attractions of each area.

  b) Which area do you think would be most appealing to each of the following types of tourist and why

    ● a family with young children

    ● a group of GAP year students who have just left university

    ● an active retired couple.

  c) Which area would you most like to visit and why?

3 Study Figure 3.

  a) In what direction are most of the easyJet flights from London Gatwick?

  b) How many UK cities can you fly to from Gatwick?

  c) What is the only route to the Baltic States?

  d) Many of the destinations in southern Europe are to the Mediterranean. Why do you think this is?

  e) Several flights go to cities in the Alps. During what time of year do you think these flights are popular and why?

  f) Does easyJet fly to places outside Europe?

  g) easyJet is often starting to operate new routes. Suggest a new route from London Gatwick to somewhere in Europe and give reasons for your choice.

RESEARCH

Plan a visit to an area of your choice in Europe served by easyJet. Use Figure 3 to give you some ideas, although these routes will have now been updated.

1) Access the easyJet website at **www.easyjet.co.uk**. Choose to fly from London Gatwick or from your closest regional airport.

2) Select two or three possible destinations from the menu and conduct some initial research on the internet to enable you to select your preferred destination.

3) For your chosen destination, write a short summary outlining why you wish to visit this place/area. Identify possible dates of travel during a forthcoming holiday and suggest flight times. How much do the flights cost?

4) You could extend this research by suggesting an itinerary and also possible accommodation.

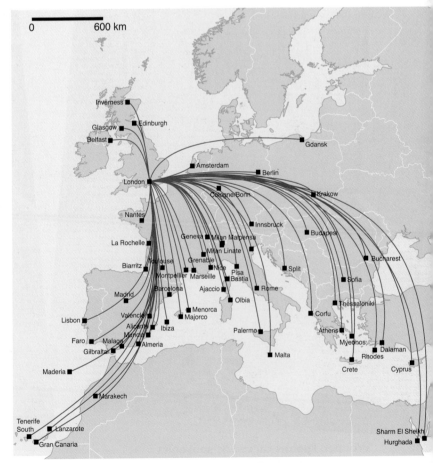

▲ **Figure 3** **easyJet from London Gatwick to Europe (2008)**

B Tourism: the good, the bad and the ugly

Tourism can bring great benefits to local communities. New jobs are created working in hotels and restaurants or acting as guides. New buildings will provide local people with jobs in the construction industry. Local farmers will have a bigger market for their fresh fruit and vegetables. Shops and local craftsmen will benefit from more trade (Figure 4). Improvements to roads and other services will benefit local people as well as the tourists. You can see why many places are keen to promote tourism!

However, too much tourism can be a bad thing. Areas can suffer from having too many people. They can become dirty, noisy and polluted (Figure 5). Footpaths can suffer from erosion, and important historical sites can be damaged by thoughtless behaviour (Figure 7). Local people may feel threatened or have their cultures and beliefs questioned. They may be inconvenienced by higher prices in the shops and more traffic congestion.

Tourism needs to be managed in a sustainable way to ensure that areas remain attractive in the future. It is important also to ensure that local people and communities continue to enjoy the benefit of tourism without suffering from the potential problems.

Wooden spoons carved as souvenirs for tourists in Greece

Pottery made in Italy

▲ Figure 4 Local craftsmen in European tourist area

Activities

4 Look back to the photos in Figure 2, page 120. Each area is attractive to tourism but could easily be spoiled or harmed by tourists.

   a) Select one of the photos **A** to **C** and suggest how tourism could cause harm or damage the area.

   b) What effect would this damage have on future tourism to your chosen area?

   c) Why is it important that tourism is carefully managed?

   d) For your chosen area, suggest ways that it could be managed sustainably.

5 Study Figure 5.

   a) Make a list of the disadvantages/problems associated with tourism shown in the photograph.

   b) What do you think is the main cause of these problems?

   c) Suggest ways of reducing the negative impacts of tourism in this area.

6 Design a poster (or Powerpoint presentation) to be displayed in your classroom showing the advantages and disadvantages of tourism to an area. Think about an effective design before starting your poster. For example, you could split your paper into two halves, or you could design your poster in the form of a balance (Figure 6). Include both writing and illustrations (drawings or photos from the internet). Your poster should be attractive and informative. Don't forget a title!

▲ **Figure 5 Negatives of tourism**

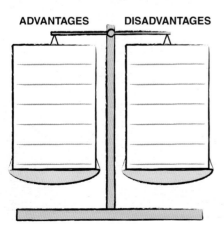

▲ **Figure 6 Balance design for poster**

▲ **Figure 7 Deliberate damage to a monument in Moscow**

C Theme Park: Disneyland Resort Paris

Theme Parks have become very popular tourist attractions in recent years. There are several huge theme parks in Europe (Figure 8) and between them they attract many millions of visitors every year.

Theme Parks, as the name suggests, are huge adventure playgrounds based on a particular theme. Whilst they may appear to be different from one another, they do have a lot in common. They all have high-tech, high-adrenalin attractions, numerous fast-food outlets and, of course, long queues! They involve a huge investment of money and there is a lot of competition between them, to offer the biggest and the fastest and the scariest rides.

It is important to understand the geography of theme parks. They can't just be located anywhere! Theme parks are massive and so they need plenty of space, both for the attractions and for the car parks and hotels. They need to be close to large population centres to attract day visitors throughout the year. They also need to have excellent transport facilities. Many theme parks try to discourage people travelling by car, in favour of using **mass transit systems** (like trains).

Disneyland Paris is one of the most popular theme parks in Europe. It opened in April 1992 and features the characters created by Walt Disney, such as Mickey Mouse and Donald Duck.

Disneyland Paris is one of two theme parks located in Disneyland Resort Paris (Figure 9) on the eastern outskirts of the French capital. The second theme park, Walt Disney Studios Park, was opened in 2002. In 2007, 15 million people visited Disneyland Paris.

Disneyland Resort Paris has excellent road and rail links, including Eurostar from the UK, and it is only a short distance from the Charles de Gaulle airport. It occupies a large area of flat land in the valley of the River Marne (Figure 10, page 126). To take account of the cold and wet European climate, there are covered walkways and more indoor features than the American equivalents. Hotels have fireplaces to keep people warm in the winter!

Siam Park, Tenerife

▲ Figure 8 European theme parks

Legoland, Denmark

RESEARCH

Select one of the other major theme parks in Europe, for example Siam Park on the Spanish island of Tenerife (www.siamparktenerife.com). Find out about the geography of the theme park (not just the rides it has to offer!). Find a location map of your chosen park and consider why it has been located where it is. Are there nearby towns? Are there good transport connections? What is the land like where the park is sited? Is there room for expansion? By all means, take some time to enjoy looking at the rides on offer, but this study is about the geography of your chosen theme park.

The following websites provide links to European theme parks:

www.themeparktravel.co.uk/europa_park_information.htm

www.dmoz.org/Recreation/Theme_Parks/Individual_Parks

▲ **Figure 9   Disneyland Paris**

Activities

7 Study Figure 10. Locate Disneyland Resort Paris on the map. Notice that it lies within the perfectly circular yellow road. This is evidence that the resort is a modern planned part of the landscape.

a) What is the number of the circular yellow road?

b) Use the numbers in red (kilometres) between the red marker pins to work out the total distance around the circular road.

c) Why do you think planners created a circular road around the theme parks?

d) Why do you think there are no main roads running directly through the parks?

e) Notice that the main railway (shown by a broken black line) runs underground below the resort. Why do you think it was designed to do this?

f) Locate the motorway (A 4-E 50) to the south of the resort. At what junction would you leave the motorway to gain direct access to the resort?

g) Having left the motorway at this junction, how many kilometres would you have to drive before reaching the car park?

8 Study Figure 10. Draw a sketch map of Disneyland Resort Paris to show its location and its main features.

a) At the centre of your sketch, draw the circular ring road. Locate the two parks, the line of the railway, the main road into the resort and the car park at its end.

b) Add to your sketch the main (red) roads around the resort. Include the motorway. Don't worry about all the minor (yellow) roads.

c) Add a scale and a north arrow. Don't forget a title too.

d) Add the following labels describing the geography of the resort:
- ring road to provide road access to the resort
- large car park
- spur road giving direct access to the resort from the motorway
- large area of flat land to provide the space needed for the resort
- good rail transport to the resort
- good road transport to the resort.

9 The owners of Disneyland Resort Paris have decided to build a third park.

a) Assuming that it has to be located within the ring road on Figure 10, where do you suggest it should be sited? Draw a simple sketch map to show your choice of site.

b) Write a couple of sentences supporting your choice of site.

c) Can you suggest a theme for the new park?

d) Describe some of the features that you would have in the park. Use diagrams and sketches to support your answer.

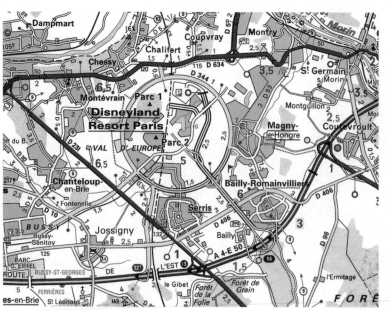

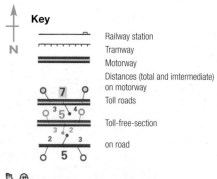

Key
- Railway station
- Tramway
- Motorway
- Distances (total and imtermediate) on motorway
- Toll roads
- Toll-free-section
- on road

MICHELIN
A better way forward

© Michelin et Cie, 2009, Authorisation No. GB0904003 Extract from ZOOM map 106 – Environs of Paris

0   2   4km

▲ Figure 10   Disneyland Resort Paris (Michelin 106 1:100,000)

D Agri-tourism on Gozo

Have you ever heard of the word **archipelago**? It is a geographical term used to describe a group of islands. In the Mediterranean Sea, the Maltese Islands are an example of an archipelago. The largest island is Malta. Gozo is the second-largest island.

Situated in the Mediterranean (Figure 11), Gozo has long been a destination for tourists wishing to enjoy the hot sunny climate and the beaches. However, recently beach tourism has declined and the island has begun to promote a new brand of tourism – **agri-tourism**.

Gozo has a beautiful and largely unspoilt rural landscape with many family-run farms. The landscape is a patchwork of fields and terraces growing fruit, vines (grapes for wine) and vegetables (Figure 12). Rural traditions, such as crafts and cooking, are strong.

Nowadays there is an increasing demand from tourists to experience local traditions and cultures. Many tourists wish to immerse themselves in real places with local people, rather than lazing around in modern concrete tourist centres surrounded by other tourists. People like to stay on farms, enjoy home cooking, witness traditional skills, and walk or cycle along unspoilt country lanes. Whilst on holiday, they want to enjoy a slower pace of life and get to know an area.

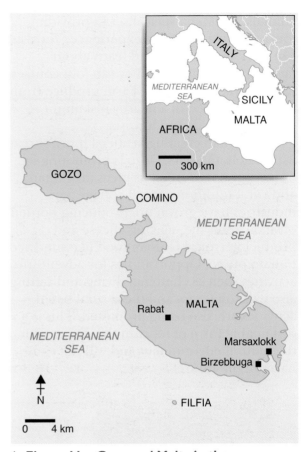

▲ **Figure 11 Gozo and Malta in the Mediterranean Sea**

▲ **Figure 12 The rural landscape of Gozo**

Activity

10 Study Figure 12.

   a) Describe the landscape in Gozo.

   b) Why is this a 'rural' landscape rather than an 'urban' landscape?

   c) What is the evidence that the area shown in the photo has a hot and dry climate?

   d) What is meant by the term 'agri-tourism'?

   e) Why do you think tourists are keen to stay on farms rather than in concrete hotels in modern tourist developments?

   f) If you and your family stayed on a farm in Gozo, what would you be interested to do?

   g) Why is agri-tourism a sustainable form of tourism?

The first of Gozo's agri-tourism enterprises was the Ta' Mena Estate (Figure 13). A mixed farm with vines, olives, fruit, vegetables and animals, Ta' Mena specialises in providing tourists with a real farm experience. Tourists can work on the farm and learn about farming techniques, or they can join courses, for example on cookery. Farm produce from the estate is used in a nearby restaurant, enjoyed by locals and tourists.

Rather than creating artificial tourist areas with hotels and bars, Gozo is promoting its own natural resources, its pleasant climate and rural traditions. Small cottage agri-industries have grown up, producing bottled olive oil and wine, making cheese and producing typical Gozoan food like sundried tomato puree. Opportunities for adventurous activities such as climbing, diving and rafting also exist on Gozo, and these rural-based activities are also being promoted. This is a sustainable form of tourism that fits in with the natural environment and will not cause harm in the future.

▲ **Figure 13 Ta' Mena Estate**

RESEARCH

The Ager Foundation is a non-profit making organisation that promotes rural tourism in Gozo. It promotes a wide range of fascinating activities based in rural areas, including birdwatching, cheese making, being a shepherd for the day, and spending a day on a traditional fishing boat. Find out more about some of the tourist activities that the organisation has to offer.

- Access the Ager Foundation's website at www.agerfoundation.com

- Make a list of some of the rural tourist activities promoted by the organisation.

- Select one of the activities that sounds interesting and click on it to find out more. Write a few sentences describing the activity. Include a photograph if you can.

- To which groups of people do you think your chosen activity will appeal?

- Now design an advert suitable for an English magazine to promote your chosen activity. It must be a single side of A4 only and should be interesting and appealing to the audience you have identified earlier. Don't forget to include a catchy title.

Activity

11 Study Figure 13. It shows part of the Ta' Mena Estate.

a) Can you identify some of the crops being grown on the Ta' Mena Estate?

b) Notice that the land has been terraced to produce a number of flat fields. Draw a simple diagram to show this.

c) Why do you think this land has been terraced?

d) What evidence is there to suggest that the climate at this time of year is dry?

e) How do you think the view in the photograph might be different in 50 years time? Explain your answer.

E Issue: Should Maro be developed as a mass tourism resort?

The Mediterranean coast of Spain has been a popular tourist destination for many years (Figure 15). Blessed with a hot sunny climate and long stretches of sandy beaches, the Mediterranean coastline has seen rapid and large-scale developments. In the 1970s and 1980s, demand soared as northern Europeans searched for cheap package holidays with guaranteed sunshine. Almost overnight, the once peaceful coastline with its traditional fishing villages became a modern concrete strip of hotels, amusement parks and shopping malls (Figure 14). The vast expanses of open sand became overcrowded, noisy and polluted as people flocked in the thousands to the Spanish Riviera.

The development of beach tourism in Spain is a good example of **mass tourism**. Wasteful of energy and water, making limited use of local skills and focused entirely on beaches, it is a good example of unsustainable tourism.

Look at Figure 16 on page 130. It is a checklist for sustainable tourism. Very few of the boxes could be ticked for the mass beach tourist developments that took place in Spain in the 1970s and 1980s.

▼ **Figure 14   Spanish coast around Malaga**

▲ **Figure 15   Spanish coast**

Recently developments have been focused on the coastline to the east of Malaga (Figure 14 on page 129). Just to the east of Nerja is the pretty fishing village of Maro (Figure 17). Currently the village has a broad economy with fishing and farming as well as tourism. There are a small number of locally owned hotels and cottages for hire. Local produce is used in the hotels and restaurants and the buildings make use of local stone. There are small-scale local industries including one specialising in solar energy, which has attracted much interest in the village. Maro is a close-knit supportive community keen to reduce waste and prevent pollution. It is a good example of a community with **sustainable tourism**. See how many boxes it ticks in Figure 16.

However, all is not well in Maro. A large-scale development is planned along the beautiful, largely unspoilt coast, involving newly built holiday and retirement homes, together with a number of road transport developments to cope with an expected increase in cars. This development, promoted by a powerful local landowner, threatens to destroy the character of the village. The local people are against the development and have staged protests. Should the development be allowed to go ahead?

Indicator		Sustainability characteristics
Energy		Maximise energy efficiency
		Use renewable sources of energy
Waste		Reduce waste
		Encourage re-use of waste materials
		Encourage recycling
Water		Use water efficiently
		Re-use waste water
Pollution		Reduce pollution (noise, air, water, land)
Buildings		Conserve and make use of older buildings
		Use local natural building materials rather than concrete
Wildlife/ environment		Conserve natural habitats
Local community		Involve local people in work, provision of skills and food produce
		Involve local community in development projects
		Ensure local communities benefit from tourism

▲ **Figure 16 Sustainable tourism checklist**

Activity

12 The aim of this activity is to develop a campaign to prevent the proposed development in Maro. It could be called the 'Save Maro Campaign', or you might think of a better heading. Your campaign could take the form of one of the following:

● a poster
● a newspaper advertisement
● a radio broadcast
● a short video
● a Powerpoint presentation.

The central thrust of your campaign is that Maro is functioning very well as a sustainable village community. You need to stress what this means and why it is important (use Figure 16 to help you).

You should consider some of the problems that may result from the proposed large-scale developments. Try to think beyond the immediate and most obvious impacts. What will be the long-term issues that might arise? Again, refer to Figure 16 to help you make the case about sustainability.

Be unashamedly one-sided in your presentation. You are against the development.

▼ **Figure 17 Maro, Spain**

Environment

A Environmental issues in Europe

Black Sea – in 2007, a severe storm led to the break-up of a Russian oil tanker, which caused a huge oil spill; thousands of migrating seabirds were killed

Crete – a massive tourist development on the Sindero peninsula is threatening to destroy large areas of natural vegetation and wildlife

Mediterranean Sea – overfishing and coastal tourist developments are affecting coastal ecosystems and turning the Mediterranean Sea into a 'marine graveyard'

Baltic Sea – a gas pipeline from Russia could disturb discarded chemical weapons and stir up a deadly cocktail of chemicals, causing an ecological disaster

River Rhône – industrial pollution has poisoned parts of the River Rhône in France, leading to a ban on the consumption of fish as toxins enter the food chain

Southern Europe – in 2003, a large number of fires broke out across southern Europe during the height of a heatwave; large areas of forest were destroyed; many of the fires were started deliberately.

Barcelona – water shortages due to overuse have led to serious effects on fragile Mediterranean ecosystems; water now has to be imported by tanker

▲ **Figure I   European environmental 'hotspots'**

You have already discovered (in earlier chapters) that Europe is fortunate to have a great range of natural environments. In the north there are the wild and largely untouched sub-polar landscapes of Iceland, Norway and Finland. There are also large areas of forest and woodland, mountains, lakes and seas. As the population of Europe has grown and people have sought to exploit resources (such as coal and oil), so the environment has begun to suffer.

Look at Figure 1. It identifies and locates a number of environmental concerns and issues that have hit the headlines in recent years. Notice that there is a range of issues involving pollution, damage to ecosystems and the unsustainable exploitation of resources. In this chapter we will consider some of the environmental challenges that lie ahead. It is in all our interests to respect and look after the environment in which we live.

RESEARCH

Carry out a short study of a recent European environmental issue. To do this, you will first need to select a recent event that interests you. A good starting point is the BBC website or other national newspaper websites, such as the Guardian or the Daily Telegraph. A Google search will reveal lots of sources of information.

Try to discover the causes of the event and its environmental effects. Use photos and maps to illustrate your account. Could the event have been prevented or was it caused naturally? What are the lessons to be learned for the future?

Activities

1 Study Figure 1. You will need a blank outline map for this activity.

a) Use the information in Figure 1 to draw a map showing recent European environmental 'hotspots'.

b) Locate each of the areas described in Figure 1 on your map of Europe. Use the atlas maps in the inside front and back covers to help you. Use colours if you wish.

c) Add some text in your own words describing each 'hotspot'.

d) Make a list of the different types of environment affected by the recent events shown in Figure 1.

e) Add simple sketches (e.g. forest fire) or photos alongside each label to make the map look more interesting and attractive.

f) Are there any other recent environmental problems that you could add to your map?

g) Give your map a title.

2 Study Figure 1 and the map you produced in question 1.

a) In what ways can tourism create environmental problems?

b) How does the transport of energy in Europe lead to environmental damage?

c) How can the weather and climate lead to environmental problems?

d) Consider the industrial pollution that has taken place on the River Rhône. How might this have been prevented?

e) What can be done to stop people starting forest fires deliberately?

B   The Mediterranean Sea: dustbin of Europe?

The Mediterranean Sea (Figure 2) is part of the Atlantic Ocean yet it is almost entirely surrounded by land. It is connected to the Atlantic by the Strait of Gibraltar, which is only 14 km wide. With such a narrow connection to the Atlantic Ocean very little water passes in and out of the Mediterranean Sea. The average depth of the Mediterranean is 1,500m although it reaches depths of over 5,000m in the Ionian Sea (Figure 2) between Greece and Italy.

The Mediterranean coastline stretches for some 46,000km. Several major European countries border the Mediterranean Sea such as Spain, France and Italy as well as countries of the Middle East and North Africa. About 150 million people live on the coast, 110 million of whom live in cities. A further 200 million people visit the region every year to enjoy the sandy beaches and warm seas.

The Mediterranean Sea is a very rich and diverse ecosystem, with many types of plant, animal and fish (Figure 3). The wellbeing of the people in the region depends on the health of the Mediterranean Sea to provide them with food and to draw tourists to the area. In recent decades people have become increasingly concerned about the impact of human activities on the fragile ecosystems and habitats of the Mediterranean Sea.

Activity

3 Study Figure 2.
  a) How many countries border the Mediterranean Sea?
  b) What countries are islands in the Mediterranean Sea? (Be careful not to count islands that belong to countries that border the Mediterranean Sea, such as Sicily, which is part of Italy!)
  c) Which S is a Middle Eastern country bordering the Mediterranean Sea?
  d) Which L is a North African country bordering the Mediterranean Sea?
  e) Which two countries are separated by the Strait of Gibraltar?
  f) What is the name of the Sea that separates Italy from Croatia?
  g) What are the names of the two Straits that separate the Mediterranean Sea from the Black Sea?

▲ Figure 2   Mediterranean Sea

▲ **Figure 3** Rich ecosystem of Mediterranean Sea

Activity

4 Study Figure 3.
   a) What is meant by the term 'ecosystem'?
   b) What is a food chain?
   c) Which part of the ecosystem shown in the photograph is at the bottom of the food chain?
   d) Why is it important to protect the part of the ecosystem that is at the bottom of the food chain?
   e) Do you think the photograph shows a 'rich' ecosystem? Explain your answer.

What are the threats to the Mediterranean Sea?

Most of the threats to the health of the Mediterranean Sea come from the land. Several large cities discharge untreated wastewater straight into the sea. This includes raw sewage. The many major rivers that flow into the sea carry a toxic cocktail of chemicals from landfill sites, intensive farming and industries (Figure 4). On reaching the sea, the chemicals pollute the water and damage the natural ecosystems.

The rapid development of the coast for tourism has damaged coastal ecosystems and led to an increase in pollution. Hotel complexes, marinas, shopping malls and multi-lane highways have replaced the natural coastal landscapes with concrete and tarmac (look back to Figure 3 on page 55).

In the sea itself, over-fishing has led to a reduction in fish stocks (Figure 5, page 136). In 2008, concern was raised about the severe decline of Mediterranean sandbar sharks and rays. Shipping can cause pollution of the sea when leaks occur, and tourist boats can damage fragile ecosystems when they drop anchor.

▲ **Figure 4** Pollution in the Mediterranean Sea

▲ Figure 5   Mediterranean fishing boat

Activities

5 Study Figure 4 (on page 135) and Figure 5.

   a) What is the evidence that the river in Figure 7 is polluted?

   b) Some people are reluctant to eat Mediterranean shellfish. Why do you think this is so?

   c) Describe the sizes of the fish shown in Figure 5.

   d) Small fish are often caught in fishing nets. How might this lead to shortages of fish in the future?

   e) Should we be concerned that certain Mediterranean species such as the sandbar shark and rays are declining in number?

6 Study Figure 6.

   a) Make a copy of Figure 6 adding colours to make the diagram more attractive.

   b) Use the information in the text above to add detailed labels describing the threats to the Mediterranean Sea.

▲ Figure 6   Threats to the Mediterranean Sea

How can the Mediterranean Sea environment be protected?

In 1975, following concerns about the effects of pollution in the Mediterranean, 15 coastal states established the Mediterranean Action Plan (MAP). The initial focus of MAP was to reduce land-based pollution, particularly from the cities and from industries. In 2005, the United Nations Environment Programme launched 'Plan Bleu' (Figure 7) to encourage sustainable development and to help prevent further pollution of the sea. Today, 22 countries have signed up to the Action Plan. They are working together to protect the Mediterranean Sea's marine and coastal environment.

Cyprus is one of a number of countries to develop a Coastal Area Management Plan. This involves carrying out studies of water quality, land-based pollution and tourist developments with a view to improving the quality of the environment. Several beaches in Cyprus now carry the 'Blue Flag' for high water quality and it is hoped that more will do so in the future. Hotels and guesthouses strive to gain the European Union's 'Ecolabel' (Figure 8) as recognition for promoting sustainable tourism.

RESEARCH

Tourism is one of the major pressures on the Mediterranean Sea. In an attempt to encourage sustainable tourism, the European Union has introduced an environmental award called the 'Ecolabel'. Hotels and guesthouses can apply for an Ecolabel if they satisfy a number of environmental conditions: for example, conserving water and energy, using natural materials and using local food products. As tourists themselves wish to be more environmentally sensitive, so they are more likely to favour accommodation with the distinctive flower symbol (Figure 8).

Imagine that you and your family own a small guesthouse on the Mediterranean coast in Cyprus. You want to achieve the Ecolabel for your guesthouse.

1 Why do you think it might be an advantage to have an Ecolabel for your guesthouse?

2 What do you have to do in order to qualify for an Ecolabel?

3 In what ways does the Ecolabel award help to prevent further harm to the Mediterranean coast and sea?

You will find plenty of information on the internet. Here are a couple of sites to get you started.

http://ec.europa.eu/environment/ecolabel/index_en.htm

http://ec.europa.eu/environment/ecolabel/product/pg_tourism_en.htm

www.ecolabel-tourism.eu

Activity

7 Study Figure 7.
   a) Why is it important that the countries bordering the Mediterranean Sea work together to protect the sea?
   b) What do you think is meant by a 'sustainable' future for the Mediterranean Sea?
   c) If the Mediterranean Sea continues to be polluted, what effects will this have on tourism?
   d) If fewer tourists visit the region what impact will this have on the region?
   e) At the moment, many tourist boats drop anchors close to the shore. As the anchors drag along the sea bed they damage the fragile ecosystems (Figure 3). Can you suggest a practical alternative to stop this damage occurring? Draw a simple diagram to support your answer.

▲ Figure 7 Front cover of Plan Bleu

▲ Figure 8 The Ecolabel

C Europe's National Parks

A national park is most commonly an area of attractive natural landscape that is protected and conserved for future generations to enjoy. The first European national parks were designated in 1909 in Sweden. Currently there are over 350 national parks in Europe. There are 14 national parks in the UK including Dartmoor, the Peak District and Snowdonia. Can you name any others? Which is your closest national park?

There are strict planning guidelines in national parks and new developments (such as roads and industries) are strictly controlled. People are encouraged to visit and enjoy national parks and there are usually a number of education and tourist information centres within the parks.

Europe's national parks protect a great variety of landscapes and natural environments (Figure 9), including mountains, lakes, rivers, peatlands, forests and wetlands. In this chapter we will study Lahemaa National Park in Estonia. You will then have the opportunity to carry out your own research into a national park of your choice.

Activity

8 Study Figure 9.
   a) For each photograph, describe the landscape and suggest why it is worthy of being a national park.
   b) Which of the landscapes would you most like to visit and why? Consider what you might like to do in your chosen landscape.

▲ **Figure 9a   Plitvice Lakes National Park, Croatia**

▲ **Figure 9b   El Torcal, Spain**

▲ **Figure 9c   Sarek National Park, Sweden**

D Lahemaa National Park, Estonia

Lahemaa National Park is located on the north Baltic Sea coast of Estonia (Figure 10). It was designated in 1971 with the aim of conserving its unique coastal environment for future generations to enjoy. Whilst people are encouraged to visit some areas, other areas have no public access at all in order to preserve the natural environment without any human interference.

Lahemaa is an extraordinary environment of low-lying coastal plains, sand dunes and inland bogs and rivers (Figure 11). Much of it is wetland, supporting a great diversity of plants, birds and animals including beaver, moose, brown bear and lynx. For much of the year it is a place of serenity and peacefulness, with relatively few visitors.

▲ **Figure 10   Location of Lahemaa National Park, Estonia**

▲ **Figure 11   Lahemaa National Park, Estonia**

Those who choose to visit Lahemaa do so to enjoy the many walks in the area (Figure 12) or to go camping or cycling. The coastline has attractive beaches often strewn with boulders (called **erratics**) dumped by the ice sheets as they retreated several thousand years ago (Figure 13).

▲ Figure 12 Walking trail, Lahemaa National Park, Estonia

▲ Figure 13 Erratics, Lahemaa National Park, Estonia

Activities

9 Study Figure 15. It is a detailed map of the Kasmu peninsula, one of the most popular parts of Lahemaa National Park. The photographs Figure 12 and 13 were taken in the northeast corner of the peninsula.

a) Draw the symbol used to indicate large erratic boulders.

b) Most of the peninsula is covered by woodland. How is this shown on the map?

c) What types of accommodation are available in Käsmu?

d) There are several information boards in the area. How are they indicated on the map and what are their main benefits?

e) Part of the Käsmu peninsula is a Special Management Zone where people do not have free access. Do you think it is a good idea to have such areas within a national park? Explain your answer.

f) Why do you think cars are prohibited from some of the tracks?

g) Make a list of the attractions and activities available to people visiting the peninsular.

h) If you had a couple of hours in the area where would you visit and why?

i) What do you think are the main challenges facing the national park authority that has to manage the area shown in Figure 15? Give reasons for your answer.

10 Study Figure 14. It is a list of 'dos and don'ts' for Lahemaa National Park.

a) Work in pairs to suggest reasons for each of the rules.

b) Design simple signs to be placed at appropriate places within the National Park to inform visitors of each of the rules. For example, a sign showing a picked wild flower with a large red cross through it, would inform visitors not to pick wild flowers. Make your signs as clear as possible. Do not use words, as some people may not understand the language.

Park your car in a car park, not on the verge

Forest roads and paths are only for hiking or cycling

Pitch tents and have campfires only in places where this is permitted

Be sensitive to nesting birds from 1 April to 15 July, and do not touch birds' nests

Please do not pick wild flowers

Please do not make unnecessary noise that could affect the wildlife and spoil other people's enjoyment of the forests

Please take your litter home

▲ **Figure 14 Lahemaa National Park: dos and don'ts**

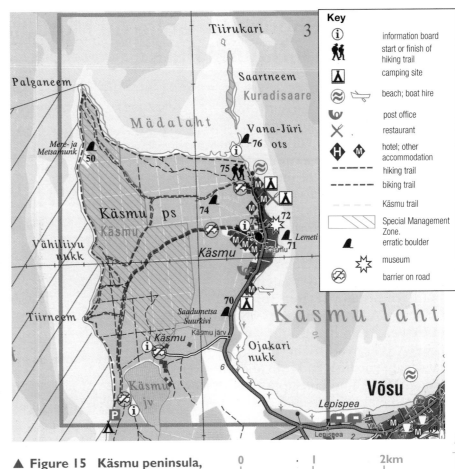

▲ **Figure 15 Käsmu peninsula, Lahemaa National Park, Estonia (1:60,000)**

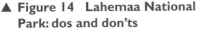

RESEARCH

Conduct your own study of a European national park. Begin by accessing the list of European national parks at http://en.wikipedia.org/wiki/List_of_national_parks #Europe. Scroll through and select a country that interests you and then take a look at the various national parks in your chosen country. Once you have chosen a national park, attempt the following:

1 Find a location map that shows where your chosen national park is located within the country. A simple Google search will reveal several options. Select a map that is clear and not too cluttered.

2 What are the special qualities of your chosen national park? Try to discover what makes it special and why it has become a national park. Why does it need special protection?

3 Why do people visit the national park? What are the attractions and what are the opportunities on offer?

4 Select one place that you would be interested to visit. Find a photograph of this place and describe why you would like to go there.

Other useful websites include the following:

http://nationalparks.wikia.com/wiki/European_National_Parks_Centre

www.visiteurope.com/ccm/experience/detail/?nav_cat=134&lang=pt_GL&item_url=/ETC/pan-european/european-national-parks/european-national-parks.pt

www.europarc.org

E Issue: Should the Golfe du Morbihan become a Regional Natural Park?

In France, there are currently nine National Parks and 50 Regional Natural Parks. A Regional Natural Park (Parc Naturel Regional) is a rural area of outstanding natural beauty. It is strictly managed to protect its scenery and historical heritage and to safeguard a sustainable future for the people living there.

The Golfe du Morbihan is a seascape of exceptional beauty located in southern Brittany (Figure 16). The turbulent Atlantic waters provide plenty of oxygen, resulting in a thriving ecology (Figure 17). There are abundant shellfish and fish, and the marshes and mudflats support a diverse range of plant species and birds.

In recent years an increase in tourism, together with a surge in the building of second homes, has put pressure on the fragile ecosystems of the Golfe. New roads are being built as well as amenities (such as shops, hotels and swimming pools). As more of the land is developed so natural habitats are being destroyed, threatening the long-term survival of the area.

Environmental organisations and local people believe that the Golfe du Morbihan should become a Parc Naturel Regional, as this status would offer protection against continued development. What do you think?

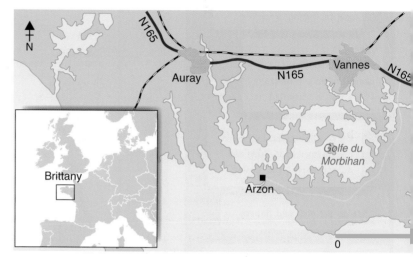

▲ **Figure 16 Golfe du Morbihan, Brittany, France**

RESEARCH

The purpose of this activity is to make a case for the Golfe du Morbihan to be designated as a Parc Naturel Regional. This would control the type and pace of development in the area and would help to sustain its unique character.

You can work individually or in groups of two or three. You can present your case in the form of a poster, a leaflet or brochure (using Publisher) or in the form of a Powerpoint presentation.

In making your case for the protection of the Golfe du Morbihan, you should use the internet to discover what makes the area special.

1 Try to find some information about the natural environment, the seashore, the mudflats and the saltmarshes. Are there any rare plants and animals that depend on the habitats of the area? Is the birdlife really special?

2 What are the pressures on the area? Have there been any new tourist developments? Can you find evidence of recent house building?

3 Finally, suggest why you think the area should become a protected Parc. You can use the first of the weblinks below to find out more about the Parc Naturel Regionals.

http://en.wikipedia.org/wiki/Regional_natural_parks_of_France

www.france-for-visitors.com/brittany/south/golfe-de-morbihan.html

▲ **Figure 17 Seascape, Gulf du Morbihan**

Ageing An increase in the proportion of older people (usually taken to be over 65) in a population

Agri-tourism Tourism based on farms, often involving rural or countryside activities such as walking, birdwatching and cooking

Arch An arch-like feature at the coast where a headland has been broken through by coastal erosion

Archipelago A group of islands, such as the many small islands off the coast of Stockholm in Sweden

Backwash The backward movement of waves as they drain back towards the sea

Bauxite The reddy-brown mineral raw material used to manufacture aluminium

Biome A global ecosystem, such as a tropical rainforest or coniferous forest

Cave An enlarged crack hollowed out in a cliff by coastal erosion

Choropleth A type of mapping where a range of increasingly dark colours is used to represent data grouped into categories

Citrus fruit Fruit such as oranges, grapefruit and lemons

Climate The long-term weather conditions averaged over a period of 30 years

Climate graph A graph showing temperature and precipitation over a period of a year for a particular location

Components Separate parts, for example, engines, tyres, wheels of a car

Corrasion A process of erosion. At the coast this involves rock fragments being picked up and flung at a cliff

Crevasse A deep crack in a glacier

Desalination A process that converts saltwater (usually from the sea) into freshwater for drinking

Distribution The spread, for example, of deserts in the world or supermarkets in a town

Ecosystem A group of living and non-living components of an environment and the interactions between them. An ecosystem can be as small as a hedge or a pond

Equator 0 degrees line of latitude. The Equator is about 40,000 km in length

Erosion The wearing away and removal of material, for example, at the coast or in a river

Erratics Isolated rocks deposited by glaciers as they retreat from a landscape

European Free Trade Association An alternative group of European countries (Iceland, Lichtenstein, Norway and Switzerland) that cooperate with each other much like the European Union

European Union A group of European countries that work together on economic, social and environmental matters. In 2009 there were 27 member of the E.U.

Fjord A very deep drowned glacial valley commonly found on the west coast of Norway

Free trade Trade between countries that is not subject to restrictive government controls

Groves The term often used to describe land devoted to fruit trees, such as oranges

Headland An area of land at the coast jutting out into the sea. A headland is often made of tough rock

Hydraulic action A process of erosion, it is the sheer power of flowing water

Inputs Aspects that form inputs to a system. Soils, climate, seeds and fertiliser are inputs to the farm system

Longshore drift The zig-zag movement of pebbles along a beach caused by waves approaching the coastline at an angle

Manufactured Manufactured goods have been made from raw materials. A car is a good example

Maquis Low growing shrubs and bushes typical of a Mediterranean biome

Mass tourism Resorts or activities that cater for large numbers of people, such as beach tourism in Spain

Mass transit systems A transport system intended to carry large numbers of people, for example by bus or train, usually in large cities

Mixed farming A mix of arable (crops) and pastoral (livestock)

GLOSSARY

Nocturnal   Another word for 'night-time'. In hot environments, many animals are nocturnal. They are active at night when conditions are cooler

Non-renewable energy   Energy that cannot be re-used, such as oil and gas

North Atlantic Drift   A warm ocean current that originates in the South Atlantic near the Caribbean and brings warm conditions to north west Europe

Off-piste   A term used to describe areas of snow and ice away from regulated ski slopes

Organic   A type of farming that does not involve the use of artificial chemicals

Outputs   Outputs from the system, such as wheat, animals and straw on a farm

Patterns   Regular, often repeating trends, such as circular or grid square. Geographers are interested in patterns of settlement or roads

Perception   A person's individual viewpoint about an issue

Permeable rock   A rock that allows water to pass through it

Primary industry   This involves extracting raw materials, such as minerals. It also included fishing, farming and forestry

Processes   Processes convert inputs into outputs. On a farm, processes include ploughing and harvesting

Renewable energy   Energy that can be re-used, such as wind and water

Secondary industry   Another term for manufacturing, so it would include car factories

Siesta   A period of rest in the afternoon traditionally taken in hot climates

Stack   An isolated pinnacle of rock formed when the roof of an arch collapses

Stump   An eroded stack, a stump is only exposed at low tide

Sustainable   Something that is long lasting and does not damage the environment. Wind and water energy are good examples of sustainable energy

Sustainable city   A city that tries to follow sustainable policies, such as its use of energy and management of waste

Sustainable tourism   Tourism that does not cause any long-term damage to the environment or the communities in which it occurs. It usually takes place in natural environments and involves small numbers of people

Swash   The forward movement of waves as they break up a beach

System   A group of component parts and the linkages that join them together, for example on a farm

Theme Parks   Large entertainment parks with a particular theme, such as Disneyland Paris

Transnational company   A company that operates and has offices and factories all over the world, for example, BMW

Weather   The day to day conditions of the atmosphere (e.g. rainfall, sunshine and temperature)